Encyclopedia of Electronic Components Volume 2

Charles Platt
with Fredrik Jansson

MAKER MEDIA

SEBASTOPOL, CA

Encyclopedia of Electronic Components Volume 2

by Charles Platt
with Fredrik Jansson

Published by Maker Media, Inc., 1160 Battery Street East, Suite 125, San Francisco, CA 94111.

Maker Media books may be purchased for educational, business, or sales promotional use. Online editions are also available for most titles (*http://oreilly.com*). For more information, contact our corporate/institutional sales department: 800-998-9938 or *corporate@oreilly.com*.

Editor: Brian Jepson
Production Editor: Melanie Yarbrough
Proofreader: Jasmine Kwityn
Indexer: Last Look Editorial

Cover Designer: Karen Montgomery
Interior Designer: David Futato
Illustrator and Photographer: Charles Platt

November 2014:   First Edition

Revision History for the First Edition

2014-11-10:   First Release
2015-01-09:   Second Release
2016-07-05:   Third Release
2017-10-20:   Fourth Release
2019-12-06:   Fifth Release

See *http://oreilly.com/catalog/errata.csp?isbn=0636920026150* for release details.

978-1-449-33418-5

[LSI]

In fond memory of my father, Maurice Platt

Table of Contents

> INTEGRATED CIRCUIT

> > ANALOG

> LIGHT SOURCE, INDICATOR, OR DISPLAY

> > REFLECTIVE

> > SINGLE SOURCE

> > MULTI-SOURCE OR PANEL

How to Use This Book

This is the second of three volumes. Its purpose is to provide an overview of the most commonly used electronic components, for reference by students, engineers, hobbyists, and instructors. While you can find much of this information dispersed among datasheets, introductory books, websites, and technical resources maintained by manufacturers, the *Encyclopedia of Electronic Components* gathers all the relevant facts in one place, properly organized and verified, including details that may be hard to find elsewhere. Each entry includes typical applications, possible substitutions, cross-references to similar devices, sample schematics, and a list of common problems and errors.

You can find a more detailed rationale for this encyclopedia in the Preface to Volume 1.

Volume Contents

Practical considerations influenced the decision to divide this encyclopedia into three volumes. Each deals with broad subject areas as follows.

Volume 1

Power; electromagnetic devices; discrete semiconductors

The *power* category includes sources of electricity and methods to distribute, store, interrupt, convert, and regulate power. The *electromag-netic devices* category includes devices that exert force linearly, and others that create a turning force. *Discrete semiconductors* include the primary types of diodes and transistors. A contents listing for Volume 1 appears in Figure P-1.

Volume 2

Thyristors (SCRs, diacs, and triacs); integrated circuits; light sources, indicators, and displays; and sound sources

Integrated circuits are divided into analog and digital components. *Light sources, indicators, and displays* are divided into reflective displays, single sources of light, and displays that emit light. *Sound sources* are divided into those that create sound, and those that reproduce sound. A contents listing for Volume 2 appears in Figure P-2.

Volume 3

Sensing devices

The field of sensors has become so extensive, they easily merit a volume to themselves. *Sensing devices* include those that detect light, sound, heat, motion, pressure, gas, humidity, orientation, electricity, proximity, force, and radiation.

At the time of writing, Volume 3 is still in preparation, while Volume 1 is complete and is available in a variety of formats.

Primary Category	Secondary Category	Component Type
power	source	battery
	connection	jumper
		fuse
		pushbutton
		switch
		rotary switch
		rotational encoder
	moderation	relay
		resistor
		potentiometer
		capacitor
		variable capacitor
	conversion	inductor
		AC-AC transformer
		AC-DC power supply
		DC-DC converter
		DC-AC inverter
	regulation	voltage regulator
electro-magnetism	linear output	electromagnet
		solenoid
	rotational output	DC motor
		AC motor
		servo motor
		stepper motor
discrete semi-conductor	single junction	diode
		unijunction transistor
	multi-junction	bipolar transistor
		field-effect transistor

Figure P-1 *The subject-oriented organization of cate-gories and entries in Volume 1.*

Primary Category	Secondary Category	Component Type
discrete semi-conductor	thyristor	SCR
		diac
		triac
integrated circuit	analog	solid-state relay
		optocoupler
		comparator
		op-amp
		digital potentiometer
		timer
	digital	logic gate
		flip-flop
		shift register
		counter
		encoder
		decoder
		multiplexer
light source, indicator or display	reflective	LCD
	single source	incandescent lamp
		neon bulb
		fluorescent light
		laser
		LED indicator
		LED area lighting
	multi-source or panel	LED display
		vacuum-fluorescent
		electroluminescence
sound source	audio alert	transducer
		audio indicator
	reproducer	headphone
		speaker

Figure P-2 *The subject-oriented organization of cate-gories and entries in Volume 2.*

Organization

Reference versus Tutorial

As its title suggests, this is a reference book, not a tutorial. A tutorial begins with elementary concepts and builds sequentially toward concepts that are more advanced. A reference book assumes that you may dip into the text at any point, learn what you need to know, and then put the book aside. If you choose to read it straight through from beginning to end, you will find some repetition, as each entry is intended to be self-sufficient, requiring minimal reference to other entries.

My books *Make: Electronics* and *Make: More Electronics* follow a tutorial approach. They don't go into as much depth as this Encyclopedia, because a tutorial inevitably allocates a lot of space to step-by-step explanations and instructions.

Theory and Practice

This book is oriented toward practicality rather than theory. I assume that the reader mostly wants to know how to use electronic components, rather than why they work the way they do. Consequently, I have not included proofs of formulae or definitions rooted in electrical theory. Units are defined only to the extent necessary to avoid confusion.

Many books on electronics theory already exist, if theory is of interest to you.

Entries

This encyclopedia is divided into entries, each entry being devoted to one broad type of component. Two rules determine whether a component has an entry all to itself, or is subsumed into another entry:

Rule 1

A component merits its own entry if it is (a) widely used, or (b) not so widely used but has a unique identity and maybe some historical status. The **bipolar transistor** entry is an example of a widely used component, whereas the **unijunction transistor** entry is an example of a not so widely used component with a unique identity.

Rule 2

A component does not merit its own entry if it is (a) seldom used, or (b) very similar in function to another component that is more widely used. For example, a *rheostat* is subsumed into the **potentiometer** section, while *silicon diode*, *Zener diode*, and *germanium diode* are combined together in the **diode** entry.

Inevitably, these guidelines required judgment calls which in some cases may seem arbitrary. My ultimate decision was based on where I would expect to find a component if I was looking for it myself.

Subject Paths

Entries are not organized alphabetically. They are grouped by subject, in much the same way that books in the nonfiction section of some libraries are organized by the Dewey Decimal System. This is convenient if you don't know exactly what you are looking for, or if you don't know all the options that may be available to perform a task that you have in mind.

Each primary category is divided into subcategories, and the subcategories are divided into component types. This hierarchy is shown in Figure P-2. It is also apparent when you look at the top of the first page of each entry, where you will find the path that leads to it. The **diac** entry, for instance, is headed with this path:

discrete semiconductor > thyristor > diac

Any classification scheme will run into exceptions. You can buy a chip containing a *resistor array*, for instance. Technically, this is an *analog integrated circuit*, but a decision was made to put it in the **resistor** section of Volume 1, because it can be directly substituted for a group of resistors.

Some components have hybrid functions. A **multiplexer**, for instance, may pass analog signals and may have "analog" in its name. However, it is digitally controlled and is mostly used in conjunction with other digital integrated circuits. This seemed to justify placing it in the digital category.

Inclusions and Exclusions

There is also the question of what is, and is not, a component. Is wire a component? Not for the purposes of this encyclopedia. How about a **DC-DC converter**? Because converters are now sold in small packages by component suppliers, they are included in Volume 1 as components.

Many similar decisions had to be made on a case-by-case basis. Some readers will disagree with the outcome, but reconciling all the disagreements would have been impossible. The best I could do was to create a book which is organized in the way that would suit me best if I were using it myself.

Typographical Conventions

Within each entry, **bold type** is used for the first occurrence of the name of a component that has its own entry elsewhere. Other important electronics terms or component names may be presented in *italics*.

The names of components, and the categories to which they belong, are all set in lowercase type, except where a term is normally capitalized because it is an acronym or a trademark. The term *Trimpot*, for instance, is trademarked by Bourns, but *trimmer* is not. **LED** is an acronym, but *cap* (abbreviation for **capacitor**) is not.

The European convention for representing fractional component values eliminates decimal points. Thus, values such as 3.3K and 4.7K are expressed as 3K3 and 4K7. This style has not been adopted to a significant degree in the United States, and is not used in this encyclopedia.

In mathematical formulae, I have used the style that is common in programming languages. The

* (asterisk) is used as a multiplication symbol, while the / (forward slash) is used as a division symbol. Where some terms are in parentheses, they must be dealt with first. Where parentheses are inside parentheses, the innermost ones must be dealt with first. So, in this example:

$$A = 30 \; / \; (7 + (4 * 2) \;)$$

You would begin by multiplying 4 times 2, to get 8; then add 7, to get 15; then divide that into 30, to get the value for A, which is 2.

Visual Conventions

Figure P-3 shows the conventions that are used in the schematics in this book. A black dot always indicates a connection, except that to minimize ambiguity, the configuration at top right is avoided, and the configuration at top center is used instead. Conductors that cross each other without a black dot do not make a connection. The styles at bottom right are sometimes seen elsewhere, but are not used here.

All the schematics are formatted with pale blue backgrounds. This enables components such as switches, transistors, and LEDs to be highlighted in white, drawing attention to them and clarifying the boundary of the component. The white areas have no other meaning.

Photographic Backgrounds

All photographs of components include a background grid that is divided into squares measuring 0.1". Although the grid is virtual, it is equivalent in scale to physical graph paper placed immediately behind the component. If the component is photographed at an angle, the grid may be reproduced at a similar angle, creating perspective on the squares.

Background colors in photographs were chosen for contrast with the colors of the components, or for visual variety. They have no other significance.

Component Availability

Because there is no way of knowing if a component may have a long production run, this encyclopedia is cautious about listing specific part numbers. To find a specific part that has a narrow function, searching the websites maintained by suppliers will be necessary. The following suppliers were checked frequently during the preparation of the book:

- Mouser Electronics (*http://www.mouser.com*)

- Jameco Electronics (*http://www.jameco.com*)

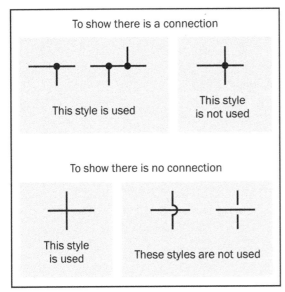

Figure P-3 *Visual conventions that are used in the schematics in this book.*

When seeking obsolete parts, or those that are nearing the end of their commercial life, eBay can be very useful.

Issues and Errata

If you believe you have found an error in this book, you will find guidance on how to report it here: *http://bit.ly/eec_v2_errata*.

Before posting your own erratum, please check those that have been submitted previously, to see if someone else already reported it.

I value and encourage reader feedback. However, before you post feedback publicly to a site such as Amazon, I have a request. Please be aware of the power that you have as a reader, and use it fairly. A single negative review can create a bigger effect than you may realize. It can certainly outweigh half-a-dozen positive reviews. If you feel you have not received a prompt or adequate response from the O'Reilly errata site mentioned here, you can email me personally at:

make.electronics@gmail.com

I check that address irregularly—sometimes only once in a couple of weeks. But I do answer all messages.

Other Information

My books *Make: Electronics* and *Make: More Electronics* are available as tutorials, using a hands-on approach with multiple projects to teach the functions of components and circuits. Component kits are available from independent suppliers. For more information, please visit *http://www.plattelectronics.com*.

O'Reilly Safari

 Safari (formerly Safari Books Online) is a membership-based training and reference platform for enterprise, government, educators, and individuals.

Members have access to thousands of books, training videos, Learning Paths, interactive tutorials, and curated playlists from over 250 publishers, including O'Reilly Media, Harvard Business Review, Prentice Hall Professional, Addison-Wesley Professional, Microsoft Press, Sams, Que, Peachpit Press, Adobe, Focal Press, Cisco Press, John Wiley & Sons, Syngress, Morgan Kaufmann, IBM Redbooks, Packt, Adobe Press, FT Press, Apress, Manning, New Riders, McGraw-Hill, Jones & Bartlett, and Course Technology, among others.

For more information, please visit *http://oreilly.com/safari*.

How to Contact Us

Please address comments and questions concerning this book to the publisher:

> Make:
> 1160 Battery Street East, Suite 125
> San Francisco, CA 94111
> 877-306-6253 (in the United States or Canada)
> 707-639-1355 (international or local)

Make: unites, inspires, informs, and entertains a growing community of resourceful people who undertake amazing projects in their backyards, basements, and garages. Make: celebrates your right to tweak, hack, and bend any technology to your will. The Make: audience continues to be a growing culture and community that believes in bettering ourselves, our environment, our educational system—our entire world. This is much more than an audience, it's a worldwide movement that Make: is leading—we call it the Maker Movement.

For more information about Make:, visit us online:

> Make: magazine: *http://makezine.com/magazine*
> Maker Faire: *http://makerfaire.com*
> Makezine.com: *http://makezine.com*
> Maker Shed: *http://makershed.com*

We have a web page for this book, where we list errata, examples, and any additional information. You can access this page at: *http://bit.ly/encyclopedia_of_electronic_components_v2*.

Acknowledgments

Any reference work draws inspiration from many sources. Datasheets and tutorials maintained by component manufacturers were considered the most trustworthy sources of information online. In addition, component retailers, college texts, crowd-sourced reference works, and hobbyist sites were used. The following books provided useful information:

Boylestad, Robert L. and Nashelsky, Louis: *Electronic Devices and Circuit Theory*, 9th edition. Pearson Education, 2006.

Braga, Newton C.: *CMOS Sourcebook*. Sams Technical Publishing, 2001.

Hoenig, Stuart A.: *How to Build and Use Electronic Devices Without Frustration, Panic, Mountains of Money, or an Engineering Degree*, 2nd edition. Little, Brown, 1980.

Horn, Delton T.: *Electronic Components*. Tab Books, 1992.

Horn, Delton T.: *Electronics Theory*, 4th edition. Tab Books, 1994.

Horowitz, Paul and Hill, Winfield: *The Art of Electronics*, 2nd edition. Cambridge University Press, 1989.

Ibrahim, Dogan: *Using LEDs, LCDs, and GLCDs in Microcontroller Projects*. John Wiley & Sons, 2012.

Kumar, A. Anand: *Fundamentals of Digital Circuits*, 2nd edition. PHI Learning, 2009.

Lancaster, Don: *TTL Cookbook*. Howard W. Sams & Co, 1974.

Lenk, Ron and Lenk, Carol: *Practical Lighting Design with LEDs*. John Wiley & Sons, 2011.

Lowe, Doug: *Electronics All-in-One for Dummies*. John Wiley & Sons, 2012.

Mims III, Forrest M.: *Getting Started in Electronics*. Master Publishing, 2000.

Mims III, Forrest M.: *Electronic Sensor Circuits & Projects*. Master Publishing, 2007.

Mims III, Forrest M.: *Timer, Op Amp, & Optelectronic Circuits and Projects*. Master Publishing, 2007.

Predko, Mike: *123 Robotics Experiments for the Evil Genius*. McGraw-Hill, 2004.

Scherz, Paul: *Practical Electronics for Inventors*, 2nd edition. McGraw-Hill, 2007.

Williams, Tim: *The Circuit Designer's Companion*, 2nd edition. Newnes, 2005.

I also made extensive use of information on vendor sites, especially:

- Mouser Electronics (*http://www.mouser.com*)

- Jameco Electronics (*http://www.jameco.com*)

- All Electronics (*http://www.allelectronics.com*)

- sparkfun (*http://www.sparkfun.com*)

- Electronic Goldmine (*http://www.goldmine-elec-products.com*)

- Adafruit (*http://www.adafruit.com*)

- Parallax, Inc. (*http://www.parallax.com*)

In addition, some individuals provided special assistance. My editor, Brian Jepson, was immensely helpful in the development of this book. Philipp Marek and Steve Conklin reviewed the text for errors. My publisher demonstrated faith in my work. Kevin Kelly unwittingly influenced me with his legendary interest in "access to tools." It was Mark Frauenfelder who originally brought me back to the pleasures of building things, and Gareth Branwyn who revived my interest in electronics.

Lastly, I should mention my school friends from decades ago: Patrick Fagg, Hugh Levinson, Graham Rogers, William Edmondson, and John Witty, who helped me to feel that it was OK to be a nerd building my own audio equipment, long before the word "nerd" actually existed.

—Charles Platt, 2014

SCR | 1

The acronym **SCR** is derived from *silicon-controlled rectifier*, which is a gate-triggered type of *thyristor*. A thyristor is defined here as a semiconductor having four or more alternating layers of p-type and n-type silicon. Because it predated integrated circuits, and in its basic form consists of a single multilayer semiconductor, a thyristor is considered to be a discrete component in this encyclopedia. When a thyristor is combined with other components in one package (as in a **solid-state relay**), it is considered to be an integrated circuit.

Other types of thyristor are the **diac** and **triac**, each of which has its own entry.

Thyristor variants that are not so widely used, such as the *gate turn-off thyristor (GTO)* and *silicon-controlled switch (SCS)*, do not have entries here.

OTHER RELATED COMPONENTS

- **diac** (see Chapter 2)
- **triac** (see Chapter 3)

What It Does

In the 1920s, the *thyratron* was a gas-filled tube that functioned as a switch and a rectifier. In 1956, General Electric introduced a solid-state version of it under the name *thyristor*. In both cases, the names were derived from the thyroid gland in the human body, which controls the rate of consumption of energy. The thyratron and, subsequently, the thyristor enabled control of large flows of current.

The **SCR** (silicon-controlled rectifier) is a type of thyristor, although the two terms are often used as if they are synonymous. Text that refers loosely to a thyristor may actually be discussing an SCR, and vice versa. In this encyclopedia, the **SCR**, **diac**, and **triac** are all considered to be variant types of thyristor.

An SCR is a solid-state switch that in many instances can pass high currents at high voltages.

Like a **bipolar transistor**, it is triggered by voltage applied to a gate. Unlike the transistor, it allows the flow of current to continue even when the gate voltage diminishes to zero.

How It Works

This component is designed to pass current in one direction only. It can be forced to conduct in the opposite direction if the reversed potential exceeds its *breakdown voltage*, but this mistreatment is likely to cause damage.

By comparison, the diac and triac are designed to be bidirectional.

The SCR has three leads, identified as anode, cathode, and gate. Two functionally identical versions of the schematic symbol are shown in Figure 1-1. Early versions sometimes included a circle drawn around them, but this style has become obsolete. Care must be taken to distinguish between the SCR symbol and the symbol that represents a **programmable unijunction transistor** (PUT), shown in Figure 1-2.

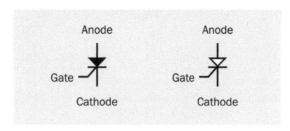

Figure 1-1. *Two functionally identical schematic symbols for an SCR (silicon-controlled rectifier). The symbol on the left is more common.*

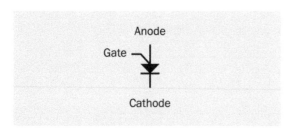

Figure 1-2. *The symbol shown here is for a programmable unijunction transistor (PUT). Care must be taken to distinguish it from the symbol for an SCR.*

Switching Behavior

When the SCR is in its passive or nonconductive state, it will block current in either direction between anode and cathode, although a very small amount of *leakage* typically occurs. When the SCR is activated by a positive voltage at the gate, current can now flow from anode to cathode, although it is still blocked from cathode to anode. When the flow reaches a level known as the *latching current*, the flow will continue even after the triggering voltage drops to zero. This behavior causes it to be known as a *regenerative* device.

If the current between anode and cathode starts to diminish while the gate voltage remains zero, the current flow will continue below the latching level until it falls below the value known as the *holding current*. The flow now ceases. Thus, the only way to end a flow of current that has been

initiated through an SCR is by reducing the flow or attempting to reverse it.

Note that the self-sustaining flow is a function of current rather than voltage.

Unlike a transistor, an SCR is either "on" or "off" and does not function as a *current amplifier*. Like a diode, it is designed to conduct current in one direction; hence the term *rectifier* in its full name. When it has been triggered, the impedance between its anode and cathode is sufficiently low that heat dissipation can be managed even at high power levels.

The ability of SCRs to pass relatively large amounts of current makes them suitable for controlling the power supplied to motors and resistive heating elements. The fast switching response also enables an SCR to interrupt and abbreviate each positive phase of an AC waveform, to reduce the average power supplied. This is known as *phase control*.

SCRs are also used to provide *overvoltage protection*.

SCR packages reflect their design for a wide range of voltages and currents. Figure 1-3 shows an SCR designed for on-state current of 4A RMS (i.e., measured as the root mean square of the alternating current). Among its applications are small-engine ignition and *crowbar* overvoltage protection, so named because it shorts a power supply directly to ground, much like a crowbar being dropped across the terminals of a car battery (but hopefully with a less dramatic outcome). See Figure 1-15.

In Figure 1-4, the SCR can handle up to 800V repetitive peak off-state voltage and 55A RMS. Possible applications include AC rectification, crowbar protection, welding, and battery charging. The component in Figure 1-5 is rated for 25A and 50V repetitive peak off-state voltage. To assess the component sizes, bear in mind that the graph line spacing is 0.1".

Figure 1-3. *SCR rated for 400V repetitive off-state voltage, no greater than 4A RMS.*

Figure 1-4. *SCR rated for 800V repetitive off-state voltage, no greater than 55A RMS.*

Internal Configuration

The function of an SCR can be imagined as being similar to that of a PNP transistor paired with an NPN transistor, as shown in Figure 1-6. In this simplified schematic, so long as zero voltage is applied to the "gate" wire, the lower (NPN) transistor remains nonconductive. Consequently, the upper (PNP) transistor cannot sink current, and this transistor also remains nonconductive.

When voltage is applied to the "gate," the lower transistor starts to sink current from the upper transistor. This switches it on. The two transistors now continue to conduct even if power to the "gate" is disconnected, because they have created a positive feedback loop.

Figure 1-5. *Stud-packaged SCR rated for 50V repetitive off-state voltage, no greater than 25A RMS.*

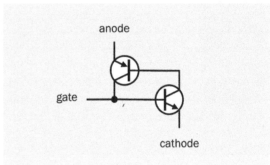

Figure 1-6. *An SCR behaves similarly to an NPN and a PNP transistor coupled together.*

Figure 1-7 shows the same two transistors in simplified form as sandwiches of p-type and n-type silicon layers (on the left), and their combination in an SCR (on the right). Although the actual configuration of silicon segments is not as simple or as linear as this diagram suggests, the SCR can be described correctly as a *PNPN device*.

An SCR is comparable with an electromagnetic **latching relay**, except that it works much faster and more reliably.

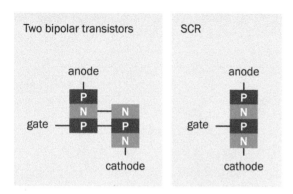

Figure 1-7. *The two transistors from the previous figure are shown here in simplified form as two stacks of p-type and n-type silicon layers. These layers are combined in an SCR, on the right.*

Breakdown and Breakover Voltage

The curves in Figure 1-8 illustrate the behavior of a hypothetical SCR, and can be compared with the curves shown for a diac in Figure 2-5 and a triac in Figure 3-10. Beginning with zero voltage applied between anode and cathode, and zero current flowing (i.e., at the center origin of the graph), if we apply a voltage at the anode that is increasingly negative relative to the cathode (i.e., we attempt to force the SCR to allow negative current flow), we see a small amount of leakage, indicated by the darker blue area (which is not drawn to scale). Finally the *breakdown voltage* is reached, at which point the negative potential overcomes the SCR and its impedance drops rapidly, allowing a surge of current to flow, probably damaging it.

Alternatively, starting once again from the center, if we apply a voltage at the anode that is increasingly positive relative to the cathode, two consequences are possible. The dashed curve assumes that there is zero voltage at the gate, and shows that some leakage occurs until the applied potential at the anode reaches the *breakover voltage*, at which point the SCR allows a large

current flow, which continues even when the voltage decreases.

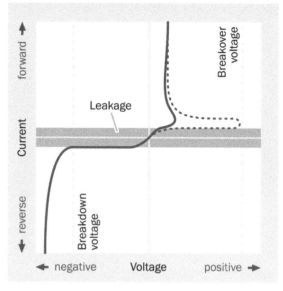

Figure 1-8. *The solid curve shows current passing between the anode and cathode of a hypothetical SCR for varying voltages, while a triggering voltage is applied to the gate. The dashed curve assumes that no triggering voltage is applied to the gate.*

In practice, the SCR is intended to respond to a positive gate voltage. Under these circumstances, its behavior is shown by the solid curve in the top-right quadrant in Figure 1-8. The SCR begins to conduct current without having to reach the breakover voltage at the anode.

- When used as it is intended, the SCR should not reach breakdown or breakover voltage levels.

SCR Concept Demo

In Figure 1-9, pushbutton S1 applies voltage to the gate of the SCR, which puts the SCR in self-sustaining conductive mode. When S1 is released, the meter will show that current continues to pass between the anode and the cathode. The X0403DF SCR suggested for this circuit has a holding current of 5mA, which a 5VDC supply should be able to provide with the 1K resistor in the circuit. If necessary, this resistor can be reduced to 680Ω.

Now if pushbutton S2 is pressed, the flow is interrupted. When S2 is released, the flow will not resume. Alternatively, if pushbutton S3 is pressed while the SCR is conducting current, the flow is diverted around the SCR, and when the pushbutton is released, the flow through the SCR will not resume. Thus, the SCR can be shut down either by a normally closed pushbutton in series with it (which will interrupt the current), or a normally open pushbutton in parallel with it (which will divert the current).

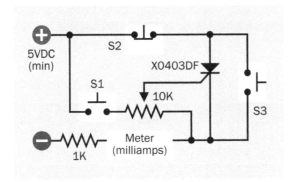

Figure 1-9. In this test circuit, S1 triggers the SCR, while S2 or S3 will stop it. See text for additional details.

The test circuit is shown installed on a breadboard in Figure 1-10. In this photograph, the red and blue wires supply a minimum of 5VDC. The two red buttons are tactile switches, the one at top left being S1 in the schematic while the one at bottom right is S3. The large switch with a rectangular button is S2; this is normally closed, and opens when pressed. The X0403DF SCR is just below it and to the right. The square blue trimmer is set to the midpoint of its range.

AC Current Applications

If the SCR is used with alternating current, it stops conducting during each negative cycle, and is retriggered in each positive cycle. This suggests one of its primary applications, as a controllable rectifier that can switch rapidly enough to limit the amount of current that passes through it during each cycle.

Figure 1-10. A breadboarded version of the SCR test circuit. The two red buttons correspond with S1 and S3 in the schematic, while the large rectangular button at top right opens S2. See text for details.

Variants

SCRs are available in surface-mount, through-hole, and stud packages, to handle increasing currents and voltages. Some special-purpose SCRs can control currents of hundreds of amps, while high-power SCRs are used to switch thousands of amps at more than 10,000V in power distribution systems. They are too specialized for inclusion in this encyclopedia.

Typical power ratings for SCRs in general use are summarized in the next section.

Values

Any SCR will impose a forward voltage drop, which typically ranges from around 1V to 2V, depending on the component.

Because SCRs are often used to modify AC waveforms, the current that the component can pass is usually expressed as the root mean square (RMS) of its peak value.

Commonly Used Abbreviations

- V_{DRM} Maximum repetitive forward voltage that may be applied to the anode while no voltage is applied to the gate (i.e., when the SCR is not in conductive mode).

- V_{RRM} Maximum repetitive reverse voltage that may be applied to the anode while no voltage is applied to the gate (i.e., when the SCR is not in conductive mode).

- V_{TM} Maximum on-state voltage while the SCR is in conductive mode. T indicates that this value changes with temperature.

- V_{GM} Forward maximum gate voltage.

- V_{GT} Minimum gate voltage required to trigger.

- V_{GD} Maximum gate voltage that will not trigger.

- I_{DRM} Peak repetitive forward blocking current (i.e., maximum leakage).

- I_{RRM} Peak repetitive reverse blocking current (i.e., leakage in the off state).

- I_{GM} Maximum forward gate current.

- $I_{T(RMS)}$ Maximum RMS current between anode and cathode while the SCR is in conductive mode. T indicates that this value changes with temperature.

- $I_{T(AV)}$ Maximum average current between anode and cathode while the SCR is in conductive mode. T indicates that this value changes with temperature.

- I_{GT} Maximum gate current required to trigger.

- I_H Typical holding current.

- I_L Maximum latching current.

- T_C Case temperature, usually expressed as an acceptable range.

- T_J Operating junction temperature, usually expressed as an acceptable range.

Surface-mount variants may tolerate maximum anode-cathode currents that typically range from 1A to 10A. Maximum voltages as high as 500V are allowed in some cases. Leakage in the "off" state may be as high as 0.5mA or as low as 5µA. Gate trigger voltage is likely to range from 0.8V to 1.5V, and trigger current of 0.2mA to 15mA is typical.

Through-hole variants may be packaged in TO-92 format (like discrete transistors) or, more commonly, in TO-220 format (like a typical 1A voltage regulator). They may be rated for a maximum of 5A up to 50A, depending on the component, with maximum voltages ranging from 50V to 500V. Leakage is similar to surface-mount variants. The gate trigger voltage is typically around 1.5V, and trigger current ranges from 25–50mA.

A stud-type SCR may have a maximum 50A to 500A current rating, although some components are capable of tolerating even higher values. Maximum voltages of 50V to 500V are possible. Leakage is likely to be higher than in other formats, with 5mA to 30mA being common. The gate trigger voltage is typically 1.5 to 3V, and trigger current may range from around 50mA to 200mA.

How to Use It

Although other applications are possible, in practice SCRs have two main applications:

- Phase control, which interrupts each positive phase of an AC power supply. It can moderate the speed of a motor or the heat generated by a resistive load.

- Overvoltage protection. This can safeguard sensitive components in a circuit where there is a DC power supply.

SCRs are often incorporated in ground-fault circuit interruptors (although not usually as discrete components) and in automotive ignition systems.

Phase Control

Phase control is a convenient way to control or limit the AC power delivered to a load by abbreviating each pulse in the AC waveform. This is done by adjusting the gate voltage so that the SCR blocks the first part of each positive phase, then conducts the remainder, and then stops conducting below its holding level. The SCR will then block the reversed flow in the negative phase of the AC waveform, but an additional SCR with opposite polarity can be added.

This is a form of *pulse-width modulation*. It is highly efficient, as the effective internal resistance of the SCR is either very high or very low, and the component does not waste significant energy in the form of heat.

On a graph showing the fluctuating voltage of an AC waveform, a single cycle is customarily divided into four stages: (1) zero voltage, (2) maximum positive voltage, (3) zero voltage, (4) minimum negative voltage, all measurements being made between the live side of the supply and the neutral side of the supply.

The cycle then repeats. Its transitions are often referred to as *phase angles* of 0 degrees, 90 degrees, 180 degrees, and 270 degrees, as shown in Figure 1-11.

The fluctuating voltage in an AC power supply is proportional with the sine of the phase angle. This concept is illustrated in Figure 1-12. If an imaginary point (shown as a purple dot) is moving in a circular path counterclockwise at a constant speed, its vertical distance (shown in green) above or below the X axis (horizontal centerline) can represent an AC voltage corresponding with the angle (shown as purple arcs) of the circle radius to the point, each angle being measured from at the center relative to a start position at right on the X axis.

When an SCR is used for phase control, the point at which it starts to conduct may be anywhere from 0 to almost 180 degrees. This is achieved by diverting a small amount of the AC power into an RC network attached to the gate of the SCR, as shown in Figure 1-13. The capacitor in this schematic introduces a delay that can be varied by the potentiometer. This enables the SCR to be triggered even after the peak of the AC power signal. In Figure 1-14, the AC power is shown by the center (green) curve, and the slightly delayed, reduced voltage at the gate is shown by the upper, purple curve. When the gate voltage rises to the trigger level, it causes the SCR to begin conducting current, creating an abbreviated output shown in the bottom curve. In this way, triggering from an AC phase angle of 0 degrees to almost 180 degrees is possible. The phase angle where the SCR begins to allow conduction is known as the *conduction angle*.

If two SCRs with opposite polarity are placed in parallel with each other, they can be used to provide phase control on both the positive-going and negative-going sections of an AC cycle. This configuration is used in high-powered devices. A **triac** is used for the same purpose with lower current.

Six SCRs may be used to control three-phase power.

Overvoltage Protection

The tolerance of an SCR for high current makes it suitable for use in a crowbar voltage limiting circuit.

In Figure 1-15, the SCR does not conduct current (other than a small amount of leakage) until the Zener diode senses a voltage above the maximum level considered safe. The diode then allows power to reach the gate of the SCR. Its impedance drops immediately, and the resulting surge of current trips the fuse. After the cause of the overvoltage condition is corrected, the fuse can be replaced and the circuit may resume functioning.

A capacitor is included so brief spikes in the power supply will be passed to ground without triggering the SCR. A resistor of around 100Ω ensures that the gate voltage of the SCR remains near zero during normal operation. When the Zener diode starts to conduct current, the resistor acts as a voltage divider with the diode, so that sufficient voltage reaches the SCR to activate it.

This circuit may be unsuitable for low-voltage power supplies, because the Zener diode has to be chosen with a high enough rating to prevent small power fluctuations from tripping it. Bearing in mind that the real triggering voltage of the diode may be at least plus-or-minus 5% of its rated voltage, the diode may have to be chosen with at least a 6V rating in a 5V circuit, and it may not be activated until the voltage is actually 6.5V. This may be insufficient to protect the components being used with the power supply.

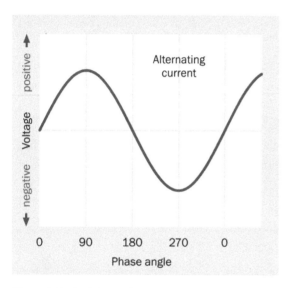

Figure 1-11. *An AC waveform is customarily measured in degrees of phase angle.*

What Can Go Wrong

Like other semiconductors, an SCR can be adversely affected by excessive heat. Usual precautions should be taken to allow sufficient ventilation and heat sinking, especially when components are moved from an open prototyping board to an enclosure in which crowding is likely.

Unexpected Triggering Caused by Heat

On a datasheet, the values for triggering current and holding current are valid only within a recommended temperature range. A buildup of heat can provoke unexpected triggering.

Unexpected Triggering Caused by Voltage

A very rapid increase in forward voltage at the anode can induce a triggering voltage in the gate by capacitive coupling. As a result, the SCR can trigger itself without any external application of gate voltage. This is sometimes known as *dv/dt triggering*. If necessary, a snubber circuit can be added across the anode input to prevent sudden voltage transitions.

Figure 1-12. *The fluctuating voltage of an AC power supply (shown as vertical green lines) is proportional with the sines of the angles (purple arcs) in this diagram. The angles are referred to as phase angles.*

Confusion of AC and DC Ratings

The on-state current for an SCR is averaged only over the width of each pulse that the SCR actually conducts. It is not time-averaged over an entire AC cycle, and it will be different again from a DC rating. Care must be taken to match the current rating with the way in which the component will actually be used.

Maximum Current versus Conduction Angle

Current-carrying capability will be very significantly affected by the length of the duty cycle when the SCR is being used to abbreviate each positive AC pulse. When the SCR imposes a 120-degree conduction angle, it may be able to handle twice the average on-state current as when it is imposing a 30-degree conduction angle. The manufacturer's datasheet should include a graphical illustration of this relationship. If an SCR is chosen for a high conduction angle, and the angle is later reduced, overheating will result, and damage is likely.

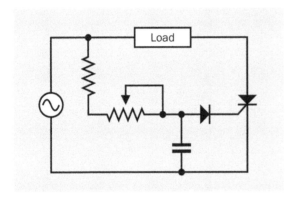

Figure 1-13. *In this schematic, an SCR is used to apply phase control, adjusting the power that passes through a load.*

Confusing Symbols

When reading a schematic, unfortunate errors can result from failure to distinguish between the symbol for a **programmable unijunction transistor (PUT)** and the symbol for an SCR. The characteristics of a PUT are described in Volume 1 of this encyclopedia.

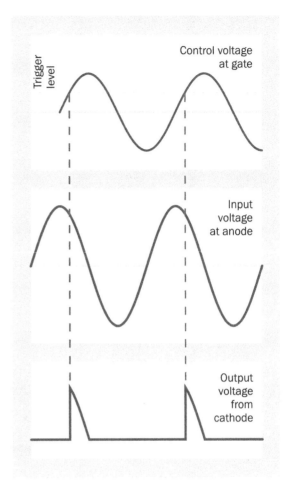

Figure 1-14. *If the AC power applied to the anode of an SCR (center) is reduced in voltage and delayed slightly by an RC network, it can trigger the SCR (top), causing it to pass only an abbreviated segment of each positive AC pulse (bottom).*

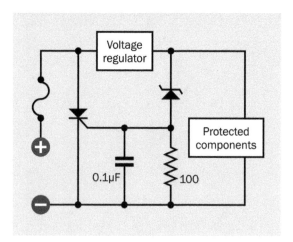

Figure 1-15. *In this schematic, an SCR is used to provide crowbar overvoltage protection for sensitive components.*

diac | 2

A **diac** is a self-triggering type of *thyristor*. Its name is said to be derived from the phrase "diode for AC," and because it is not an acronym, it is not usually capitalized.

A thyristor is defined here as a semiconductor having four or more layers of p-type and n-type silicon. Because the thyristor predated integrated circuits, and in its basic form consists of a single multilayer semiconductor, it is categorized as a discrete component in this encyclopedia. When a thyristor is combined with other components in one package (as in a **solid-state relay**) it is considered to be an integrated circuit.

Other types of thyristor are the **SCR** (silicon-controlled rectifier) and the **triac**, each of which has its own entry in this encyclopedia.

Thyristor variants that are not so widely used, such as the *gate turn-off thyristor (GTO)* and *silicon-controlled switch (SCS)*, do not have entries here.

OTHER RELATED COMPONENTS

- **SCR** (see Chapter 1)
- **triac** (see Chapter 3)

What It Does

The **diac** is a bidirectional thyristor with only two terminals. It blocks current until it is subjected to sufficient voltage, at which point its impedance drops very rapidly. It is primarily used to trigger a **triac** for purposes of moderating AC power to an **incandescent lamp**, a resistive heating element, or an AC **motor**. The two leads on a diac have identical function and are interchangeable.

By comparison, a **triac** and an **SCR** are thyristors with three leads, one of them being referred to as the gate, which determines whether the component becomes conductive. A triac and a diac allow current to flow in either direction, while an SCR always blocks current in one direction.

Symbol Variants

The schematic symbol for a diac, shown in Figure 2-1, resembles two diodes joined together, one of them inverted relative to the other. Functionally, the diac is comparable with a pair of Zener diodes, as it is intended to be driven beyond the point where it becomes saturated. Because its two leads are functionally identical, they do not require names to differentiate them. They are sometimes referred to as A1 and A2, in recognition that either of them may function as an anode; or they may be identified as MT1 and MT2, MT being an acronym for "main terminal."

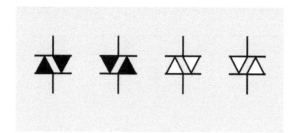

Figure 2-1. *Symbol variants to represent a diac. All four are functionally identical.*

The symbol may be reflected left to right, and the black triangles may have open centers. All of these variants mean the same thing. Occasionally the symbol has a circle around it, but this style is now rare.

When only a moderate voltage is applied (usually less than 30V) the diac remains in a passive state and will block current in either direction, although a very small amount of *leakage* typically occurs. When the voltage exceeds a threshold known as its *breakover level*, current can flow, and the diac will continue to conduct until the current falls below its *holding level*.

A sample diac is shown in Figure 2-2.

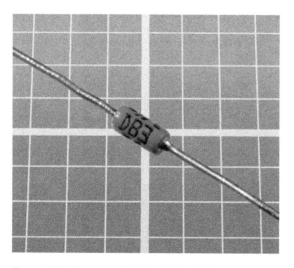

Figure 2-2. *Because a diac is not intended to pass significant current, it is typically packaged in a small format. The graph squares in the photograph each measure 0.1".*

How It Works

Figure 2-3 shows a circuit that demonstrates the conductive behavior of a diac.

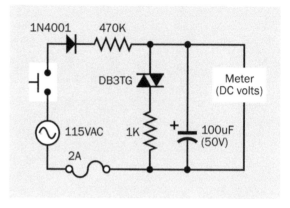

Figure 2-3. *A test circuit to demonstrate the behavior of a diac. See text for details.*

When the pushbutton is held down, current from the positive side of the AC supply flows through the diode and the 470K resistor to the capacitor. The diac is not yet conductive, so the capacitor accumulates a potential that can be monitored with the volt meter. After about 30 seconds, the charge on the capacitor reaches 32V. This is the breakover voltage for this particular diac, so it becomes conductive. The positive side of the capacitor can now discharge through the diac and the 1K series resistor to ground.

If the pushbutton is released at this moment, the meter will show that the capacitor discharges to a potential below the holding level of the diac. The capacitor now stops discharging because the diac has ceased being conductive.

If the pushbutton is held down constantly, the meter will show the capacitor charging and then discharging through the diac repeatedly, so that the circuit behaves as a *relaxation oscillator*. The 1K series resistor is included to protect the diac from excessive current. If a standard quarter-watt resistor is used, it should not become unduly warm because current passes through it only intermittently.

- Because this circuit uses 115VAC, basic precautions should be taken. The fuse should not be omitted, the capacitor should be rated for at least 50V, and the circuit should not be touched while it is connected to the power source. Breadboarding a circuit using this voltage requires caution and experience, as wires can easily come loose, and components can be touched accidentally while they are live.

Figure 2-4 shows the test circuit on a breadboard. The red and blue leads at the top of the photograph are from a fused 115VAC power supply. The live side of the supply passes through a diode to a pushbutton switch that has a rectangular black cap. A 470K resistor connects the other side of the switch to the positive side of a 100µF electrolytic capacitor, and also to the diac (small blue component). A 1K resistor connects the other end of the diac back to the negative side of the capacitor, which is grounded. The yellow and blue wires leaving the photograph at the left are connected with a volt meter, which is not shown.

Figure 2-4. *A breadboarded version of the diac test circuit. See text for details.*

The behavior of a diac is also illustrated in Figure 2-5, which can be compared with the curves in Figures 3-10 and 1-8, depicting the behavior of a triac and an SCR respectively.

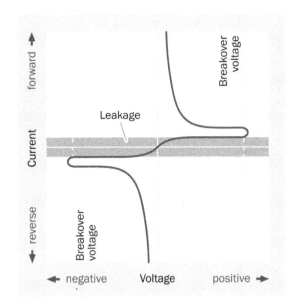

Figure 2-5. *The curve shows current passing through a diac when various voltages are applied.*

Switching AC

The diac cannot function as a switch, because it lacks the third terminal which is found in a triac, an SCR, or a bipolar transistor. However, it is well suited to drive the gate of a triac, because the behavior of a diac is symmetrical in response to opposite voltages, while the triac is not. If an AC voltage applied to a diac is adjusted with a potentiometer in an RC circuit, the diac will pass along a portion of each positive or negative pulse, and will delay it by a brief amount of time determined by the value of the capacitor in the RC circuit and the setting of the potentiometer. This is known as *phase control*, as it controls the *phase angle* at which the diac allows current to flow.

See Figure 3-13 for a schematic showing a diac driving a triac. See Figures 1-14 and 3-11 for graphs illustrating phase control. See "Phase Control" for a discussion of phase in AC waveforms.

Variants

Diacs are available in through-hole and surface-mount formats. Because they are not intended to handle significant current, no heat sink is included.

A *sidac* behaves very similarly to a diac, its name being derived from "silicon diode for alternating current." Its primary difference from generic diacs is that it is designed to reach its breakover voltage at a higher value, typically 120VAC or 240VAC.

Values

When performing its function to trigger a triac, a diac is unlikely to pass more than 100mA.

The breakover voltage of a diac is usually between 30V and 40V, with a few versions designed for up to 70V. When the diac starts to conduct, its on-state impedance is sufficient to reduce the voltage significantly, with 5V being a typical minimum output voltage.

Although the rise time when a diac responds is very brief (around 1µs), the component is not expected to run at a high frequency. It will normally be used with 50Hz or 60Hz AC to trigger a triac. For this reason, its repetitive peak on-state current is usually specified at no more than 120Hz.

Abbreviations in datasheets are likely to include:

- V_{BO} Breakover voltage (sometimes may be specified as latching voltage, which for a diac is the same thing).

- V_{BO1} - V_{BO2} Breakover voltage symmetry. The hyphen is intended as a minus sign, so that this value is the maximum difference between breakover voltage in each direction.

- V_O Minimum output voltage.

- I_{TRM} Repetitive peak on-state current.

- I_{BO} Breakover current, usually the maximum required, and less than 20µA.

- I_R Maximum leakage current, usually less than 20µA.

- T_J Operating junction temperature, usually expressed as an acceptable range.

What Can Go Wrong

Like other semiconductors, a diac is heat sensitive. Usual precautions should be taken to allow sufficient ventilation and heat sinking, especially when components are moved from an open prototyping board to an enclosure in which crowding is likely.

Unexpected Triggering Caused by Heat

On a datasheet, a value for breakover current is valid only within a recommended temperature range. A buildup of heat can provoke unexpected triggering.

Low-Temperature Effects

A higher breakover voltage will be required by a diac operating at low temperatures, although the variation is unlikely to be greater than plus-or-minus 2% within a normal operating range. Temperature has a much more significant effect on a triac.

Manufacturing Tolerances

The breakover voltage for a diac is not adjustable, and may vary significantly between samples of the component that are supposed to be identical. The diac is not intended to be used as a precision component. In addition, while its breakover voltage should be the same in either direction, a difference of plus-or-minus 2% is possible (1% in some components).

triac

3

A **triac** is a gate-triggered type of *thyristor*. Its name was probably derived from the phrase "triode for AC," and because it is not an acronym, it is not usually capitalized.

A thyristor is defined here as a semiconductor having four or more layers of p-type and n-type silicon. Because the thyristor predated integrated circuits, and in its basic form consists of a single multilayer semiconductor, it is categorized as a discrete component in this encyclopedia. When a thyristor is combined with other components in one package (as in a **solid-state relay**) it is considered to be an integrated circuit.

Other types of thyristor are the **SCR** (silicon-controlled rectifier) and the **diac**, each of which has its own entry in this encyclopedia.

Thyristor variants that are not so widely used, such as the *gate turn-off thyristor (GTO)* and *silicon-controlled switch (SCS)*, do not have entries here.

OTHER RELATED COMPONENTS

- **SCR** (see Chapter 1)
- **diac** (see Chapter 2)

What It Does

The **triac** is ubiquitous in AC dimmers for **incandescent lamps**. It is also used to control the speed of AC motors and the output of resistive heating elements. It is a type of *thyristor* which contains five segments of p-type and n-type silicon and has three leads, one of them attached to a gate that can switch a bidirectional flow of current between the other two. Its name was originally a trademark, generally thought to be derived from the phrase "triode for AC." A triode was a common type of vacuum tube when thyristors were first introduced in the 1950s.

By comparison, a **diac** is a thyristor with only two leads, allowing current to flow in either direction when the component reaches a *breakover voltage*. Its name was probably derived from the phrase "diode for AC." It is often used in conjunction with a triac.

An **SCR** (silicon-controlled rectifier) is a thyristor that resembles a triac, as it has three leads, one of them a gate. However, it only allows current to flow in one direction.

Symbol Variants

The schematic symbol for a triac, shown in Figure 3-1, resembles two diodes joined together, one of them inverted relative to the other. While a triac does not actually consist of two diodes, it is functionally similar, and can pass current in either direction.

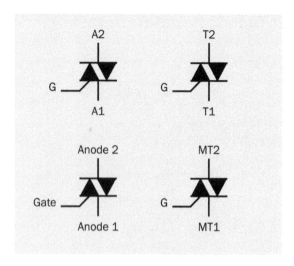

Figure 3-1. *The schematic symbol for a triac, with four naming conventions that are used for its leads. The different conventions do not indicate any functional difference.*

An appended bent line represents the gate. The labels for the other two leads are not standardized, and can be referred to as A1 and A2 (for Anode 1 and Anode 2), or T1 and T2 (for Terminal 1 and Terminal 2), or MT1 and MT2 (for Main Terminal 1 and Main Terminal 2). The choice of terms does not indicate any functional difference. In this encyclopedia entry, A1 and A2 are used.

The A1 terminal (or T1, or MT1) is always shown closer to the gate than A2 (or T2, or MT2). This distinction is important because although the triac can pass current in either direction, its behavior is somewhat asymmetrical.

- Voltages are expressed relative to terminal A1 (or T1, or MT1, if those terms are used).

The schematic symbol may be reflected or rotated, the black triangles may have open centers, and the placement of the bent line representing the gate may vary. However, terminal A1 is always nearer to the gate than terminal A2.

Figure 3-2 shows 12 of the 16 theoretical possibilities. All of these variants are functionally identical. Occasionally the symbol has a circle around it, but this style is now rare.

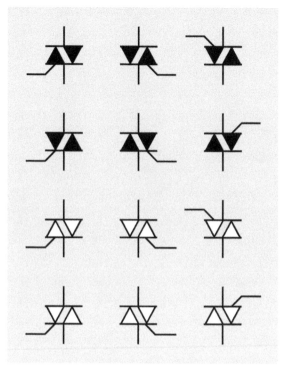

Figure 3-2. *Interchangeable variants of the schematic symbol for a triac.*

Triacs with various characteristics are shown in Figures 3-3, 3-4, and 3-5.

Figure 3-3. *The BTA208X-1000B triac can conduct 8A continuous on-state current RMS, and withstands peak off-state voltage of up to 1,000V. This is a "snubberless" triac.*

Figure 3-4. *The BTB04-600SL triac can conduct 4A continuous on-state current RMS, and withstands peak off-state voltage of up to 600V.*

Figure 3-5. *The MAC97A6 triac can conduct 0.8A continuous on-state current RMS, and withstands peak off-state voltage of up to 400V.*

How It Works

When no gate voltage is applied, the triac remains in a passive state and will block current in either direction between A1 and A2, although a very small amount of *leakage* typically occurs. If the gate potential becomes sufficiently positive *or* negative relative to terminal A1, current can begin to flow between A1 and A2 in *either* direction. This makes the triac ideal for controlling AC.

Quadrants

While a gate voltage is applied, four operating modes are possible. In each case, A1 is the reference (which can be thought of as being held at a neutral ground value). Because the triac is conducting AC, voltages above and below ground will occur. The four modes of operation are often referred to as four *quadrants*, and are typically arranged as shown in Figure 3-6.

In some reference sources (especially educational text books), current is shown with an arrow indicating a flow of electrons moving from negative to positive. Because the type of current flow is often undefined, diagrams should be interpreted carefully. In this encyclopedia, current is always shown flowing from a more-positive location to a more-negative location.

Quadrant 1 (upper right)
A2 is more positive than A1, and the gate is more positive than A1. Conventional current (positive to negative) will flow from A2 to A1. (This behavior is very similar to that of an **SCR**.)

Quadrant 2 (upper left)
A2 is more positive than A1, and the gate is more negative than A1. Once again, conventional current (positive to negative) will flow from A2 to A1.

Quadrant 3 (lower left)
A2 is more negative than A1, and the gate is more negative than A1. Conventional current is reversed from A1 to A2.

Quadrant 4 (lower right)
A2 is more negative than A1, but the gate is more positive than A1. Conventional current is reversed from A1 to A2.

- Note that two positive symbols or two negative symbols in Figure 3-6 do not mean that both locations are of equal voltage. They simply mean that these locations are at potentials that are significantly different from A1.

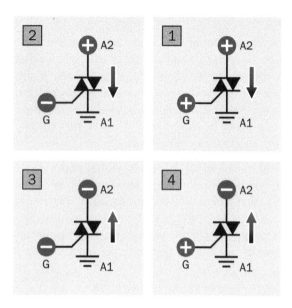

Figure 3-6. *The "quadrants" of triac behavior. Positive and negative symbols indicate which terminal is "more positive" or "more negative" than A1. The ground symbol represents a potential midway between positive and negative. See text for details.*

Suppose that gate current increases gradually. When it reaches the *gate threshold current* of the triac, the component starts conducting between A1 and A2. If the current between A1 and A2 rises above the value known as the *latching current*, it will continue to flow, even if gate current disappears completely.

If the self-sustaining current through the triac gradually diminishes, while there is no voltage applied to the gate, conduction between the main terminals will stop spontaneously when it falls below a level known as the *holding current*. This behavior is similar to that of an SCR. The triac now returns to its original state, blocking current until the gate triggers it again.

The triac is sufficiently sensitive to respond to rapid fluctuations, as in 50Hz or 60Hz AC.

Threshold, Latching, and Holding Current

Figure 3-7 shows the relationship between the gate threshold current, the latching current, and holding current. In the upper half of the figure, gate current is shown fluctuating until it crosses the threshold level. This establishes current flow between the main terminals, shown in the lower half of the figure. Prior to this moment, a very small amount of leakage current occurred (shown in the figure, but not to scale).

In this hypothetical scenario, the triac starts passing current between external components—and the current exceeds the latching level. Consequently, gate current can diminish to zero, and the triac remains conductive. However, when external factors cause the current between the main terminals to diminish below the holding level, the triac abruptly ceases to be conductive, and current falls back to the leakage level.

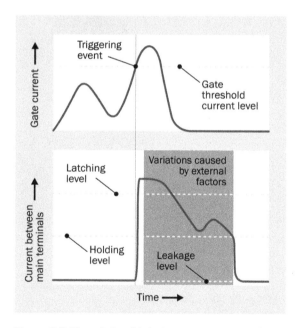

Figure 3-7. *The relationship between gate current of a triac and the current between its main terminals. See text for details.*

Unlike a **bipolar transistor**, a triac is either "on" or "off" and does not function as a *current ampli-*

fier. When it has been triggered, the impedance between A1 and A2 is low enough for heat dissipation to be manageable even at relatively high power levels.

Triac Testing

Figure 3-8 shows a circuit which can demonstrate the conductive behavior of a triac. For simplicity, this circuit is DC powered. In a real application, the triac is almost always used with AC.

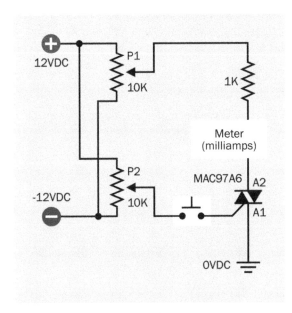

Figure 3-8. *A test circuit to show the behavior of a triac when varying positive and negative potentials are applied to the gate and to the A2 terminal, relative to A1.*

Note that this circuit requires at least a +12VDC and -12VDC power supply (higher values may also be used). The ground symbol represents a midpoint voltage of 0VDC, applied to terminal A1 of the triac, which is an MAC97A6 or similar. If a dual-voltage power supply is unavailable, the gate of the triac can be connected directly to +12VDC, omitting potentiometer P2; but in this case, only two operating modes of the triac can be demonstrated by turning potentiometer P1.

Each potentiometer functions as a voltage divider between the positive and negative sides of the power supply. P1 applies a positive or negative voltage to A2, relative to A1. P2 applies a positive or negative voltage to the gate, relative to A1.

If the test begins with both potentiometers at the top ends of their range, A1 and G both have a positive potential relative to A1, so that the triac is now in quadrant 1 of its operating modes. Pressing the pushbutton should cause it to start conducting current limited by the 1K resistor, and the meter should change from measuring 0mA to around 12mA. If the pushbutton is released, the triac should continue to conduct current, because 12mA is above this triac's latching level. If P1 is slowly moved toward the center of its range, the current diminishes, ceasing when it falls below the holding level. If P1 is now moved back to the top of its range, the current will not resume until the triac is retriggered with the pushbutton.

The test can be repeated with P1 at the top of its range and P2 at the bottom of its range, to operate the triac in quadrant 2; P1 at the bottom of its range and P2 at the bottom of its range, to operate the triac in quadrant 3; and P1 at the bottom of its range and P2 at the top of its range, to operate the triac in quadrant 4. The functionality should be the same in each case. The pushbutton will initiate a flow of current, which will diminish when P1 is turned toward the center of its range.

In any of these quadrants, P2 can be turned slowly toward the center of its range while the pushbutton is pressed repeatedly. This will allow empirical determination of the gate threshold current for this triac. The meter, measuring milliamps, will measure the current if it is inserted between the wiper of the potentiometer and the gate of the triac.

The test circuit is shown installed on a breadboard in Figure 3-9. In this photograph, the red and blue wires at left supply +12VDC and -12VDC relative to the black ground wire at top right. The yellow and green wires connect with a meter set to measure milliamps. The red button is a tactile switch, while the MAC97A6 triac is just above it and to the left. The square blue 10K trimmers are set to opposite ends of their scales, so that the meter will show current flowing when the tactile switch is pressed.

Figure 3-9. *A breadboarded triac test circuit.*

Breakover Voltage

If a much higher voltage is applied to A2, the triac can be forced to conduct current without any triggering voltage being applied to the gate. This occurs when the potential between A1 and A2 reaches the triac's *breakover voltage*, although the component is not designed to be used this way. The concept is illustrated in Figure 3-10, which can be compared with the behavior of an SCR illustrated in Figure 1-8 and the behavior of a diac shown in Figure 2-5. While the term *breakdown voltage* defines the minimum reverse voltage required to force a diode to conduct, *breakover voltage* refers to the minimum forward voltage that has this effect. Because a triac is designed to conduct in both directions, it can be thought of as having a breakover voltage in both directions.

In Figure 3-10, the numbers in yellow squares are the quadrants referred to in Figure 3-6. The solid curve represents current flow if a triggering voltage is applied to the gate while a positive or negative potential is applied to A2, relative to A1. If the gate is not triggered while the voltage be-

tween A1 and A2 gradually increases, the dashed section of the curve illustrates the outcome when the component reaches breakover voltage. Although this may not damage the triac, the component becomes uncontrollable.

- In normal usage, the voltage between A1 and A2 should not be allowed to reach breakover level.

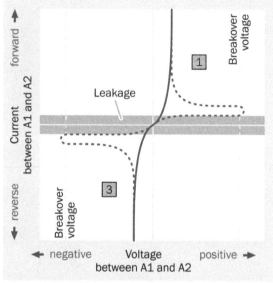

Figure 3-10. *The solid curve shows current passing between A1 and A2 in a hypothetical triac, for varying voltages, while triggering voltage is applied to the gate. The dashed curve assumes that no triggering voltage is applied to the gate. The numbers in yellow squares are the quadrants of triac operation.*

Switching AC

"Switching" AC with a triac means interrupting each pulse of current so that only a portion of it is conducted through to the load. Usually this is done with the triac functioning in quadrants 1 and 3. In quadrant 3, the polarity of the flow between A1 and A2 is opposite to that in quadrant 1, and the gate voltage is also reversed. This enables a relatively simple circuit to control the duration of each half-cycle passing through the triac. The theory of this circuit is shown in Figure 3-11.

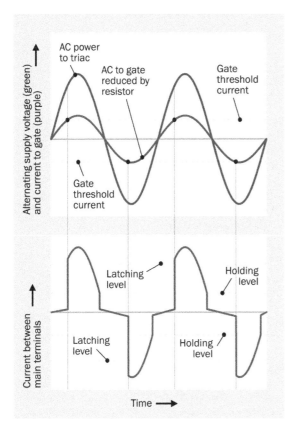

Figure 3-11. *To moderate the power of AC current, a triac can block a section of each AC pulse.*

The upper section of Figure 3-11 shows alternating voltage to the triac in green. The purple curve represents the gate current of the triac, reduced by a variable resistor. (The figure is for conceptual purposes only; the alternating power supply voltage and the fluctuating gate current cannot actually share the same vertical scale of a graph.)

Figure 3-11 can be compared with Figure 3-7, except that the negative threshold level for the gate is now shown as well as the positive threshold level. Remember, either a positive or negative voltage can activate the gate.

In Figure 3-11, initially the triac is nonconductive. As time passes, the gate current reaches the threshold level, and this triggering event enables current to flow between the main terminals of the triac, as shown in the lower part of the figure.

This current exceeds the latching level, so it continues to flow, even though the gate current diminishes below its threshold level. Finally the current between the main terminals falls below the holding level, at which point the triac stops conducting. It waits for the next triggering event, which occurs as the power supply swings to negative.

This simple system blocks a section of each AC pulse, which will vary in length depending how much current is allowed to flow through the gate. Because the blocking process occurs rapidly, we notice only the reduced overall power passing through the triac (in terms of the brightness of a light, the heat emitted by a resistive element, or the speed of a motor).

Unfortunately, there is a problem in this scenario: the triac does not quite behave symmetrically. Its gate threshold level for positive current is not exactly equal and opposite to its gate threshold level for negative current. The upper part of Figure 3-11 shows this flaw in the differing vertical offsets of the positive and negative thresholds from the central zero line.

The result is that negative AC pulses through the triac are shorter than positive pulses. This asymmetry produces harmonics and noise that can feed back into power supply wiring, interfering with other electronic equipment. The actual disparities in gate response, in each quadrant of operation for two triacs, are shown in Figure 3-12.

Current capacity of triac	Ratio of gate current required to conduct, relative to quadrant 1		
	Quadrant 2	Quadrant 3	Quadrant 4
4 amp	1.6	2.5	2.7
10 amp	1.5	1.4	3.1

Figure 3-12. *Because the internal structure of a triac is asymmetrical, it requires a different trigger current in each of its operating quadrants. This table, derived from a Littelfuse technical briefing document, shows the ratio of the minimum trigger current in quadrants 2, 3, and 4 relative to quadrant 1.*

See Figure 1-14 for a graph illustrating phase control in the SCR. See "Phase Control" for a discussion of phase in AC waveforms generally.

Triac Triggered by a Diac

The problem of asymmetrical triggering can be overcome if the triac is triggered with a voltage pulse generated by another component that does behave symmetrically. The other component is almost always a diac, which is another type of thyristor. Unlike an SCR or a triac, it has no gate. It is designed to be pushed beyond its breakover voltage, at which point it latches and will continue to conduct until current flowing through it diminishes below its holding level. See Chapter 2 for more information about the diac.

In Figure 3-13, the diac is shown to the right of the triac, and is driven by a simple RC network consisting of a fixed resistor, a potentiometer, and a capacitor. (In an actual application, the RC network may be slightly more complex.) The capacitor takes a small amount of time to charge during each half-cycle of AC. The length of this delay is adjusted by the potentiometer, and determines the point in each AC half-cycle when the voltage to the diac reaches breakover level. Because the delay affects the phase of the AC, this adjustment is known as *phase control*.

As the voltage exceeds breakover level, the diac starts to pass current through to the gate of the triac, and triggers it. The holding level of the diac is lower than its latching level, so it continues to pass current while the capacitor discharges and the voltage diminishes. When the current falls below the holding level, the diac stops conducting, ready for the next cycle. Meanwhile, the triac continues to pass current until the AC voltage dips below its holding level. At this point, the triac becomes nonconductive until it is triggered again.

This chopped waveform will still create some harmonics, which are suppressed by the coil and capacitor at the left side of the circuit in Figure 3-13.

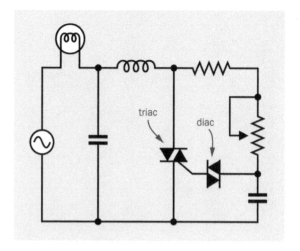

Figure 3-13. *A minimal schematic showing typical operation of a triac, with a diac supplying pulses to the triac gate. The potentiometer adjusts the delay created by the capacitor.*

Other Triac Drivers

It is possible, although unusual, to drive a triac from a source other than a diac.

Simple on-off control can be achieved by using a special optocoupler such as the MOC3162 by Fairchild Semiconductor. This emits a switching signal to a triac only when the AC voltage passes through zero. A *zero cross* circuit is desirable because it creates much less interference. The use of an optocoupler helps to isolate the triac from other components.((("zero cross circuit")))

Phase control can be achieved using an optocoupler such as the H11L1, which can be driven by rectified but unsmoothed AC after it passes through a Zener diode to limit the voltage. The output from the optocoupler is logic-compatible and can be connected with the input to a timer such as the 555, set to one-shot mode. Each pulse from the timer passes through another optocoupler such as the MOC3023, which uses an internal LED to trigger the gate of a triac.

Yet another possibility is to use the programmed output from a microcontroller, through an optocoupler, to control the gate of a triac. An online search for the terms "microcontroller" and "triac" will provide some additional suggestions.

Charge Storage

While switching AC, the internal charge between A1 and A2 inside the triac requires time to dissipate before the reverse voltage is applied; otherwise, *charge storage* occurs, and the component may start to conduct continuously. For this reason, the triac is normally restricted to relatively low frequencies such as domestic 60Hz AC power.

When a triac controls a motor, the phase lag between voltage and current associated with an inductive load can interfere with the triac's need for a transitional moment between a positive and negative voltage cycle. In a datasheet, the term *commutating dv/dt* defines the rate of rise of opposite polarity voltage that the triac can withstand without locking into a continuous-on state.

An RC *snubber network* is often wired in parallel with A1 and A2 to control the rise time of voltage to the triac, as shown within the darker blue rectangle in Figure 3-14, where a resistor and capacitor have been added just to the left of the triac. The highest resistance and lowest capacitance, consistent with trouble-free operation, should be chosen. Typical values are 47Ω to 100Ω for the resistor, and 0.01μF to 0.1μF for the capacitor.

Variants

Triacs are available in through-hole and surface-mount packages.

Some components that are referred to as triacs actually contain two SCR components of opposite polarity. The "alternistor" range from Littelfuse is an example. The SCR will tolerate faster voltage rise times than a conventional triac, and is more suitable for driving inductive loads such as large motors.

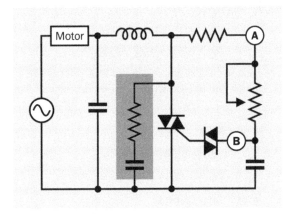

Figure 3-14. *To prevent a triac from locking itself into a continuous-on state while driving an inductive load such as a motor, a snubber circuit can be added (shown here as a resistor and capacitor in the darker blue rectangle to the left of the triac).*

A *snubberless triac*, as its name implies, is designed to drive an inductive load without need for a snubber circuit. An example is the STMicroelectronics BTA24. Datasheets for this type of component impose some limits that may be stricter than for a generic triac.

Values

Surface-mount triacs are typically rated between 2A to 25A of switched AC current (RMS), the higher-current versions being as large as 10mm square. The necessary gate trigger voltage may range from 0.7V to 1.5V. Through-hole packages may be capable of slightly higher currents (up to 40A), with gate trigger voltages of 1V to 2.5V being common.

As noted previously, the majority of triacs are restricted to relatively low frequency switching, 60Hz being very common.

Abbreviations in datasheets are likely to include:

- V_{DRM} or V_{RRM} Peak repetitive reverse off-state voltage. The maximum reverse voltage that the component will withstand in its "off" state without experiencing damage or allowing current to pass.

- V_{TM} The maximum voltage difference between A1 and A2, measured with a short pulse width and low duty cycle.

- V_{GT} Gate trigger voltage necessary to produce the gate trigger current.

- I_{DRM} Peak repetitive blocking current (i.e., maximum leakage).

- I_{GM} Maximum gate current.

- I_{GT} Minimum gate trigger current.

- I_H Holding current.

- I_L Latching current.

- $I_{T(RMS)}$ On-state RMS current. The maximum value passing through the component on a continuous basis.

- I_{TSM} Maximum non-repetitive surge current. Specified at a stated pulse width, usually 60 Hz.

- T_C Case temperature, usually expressed as an acceptable range.

- T_J Operating junction temperature, usually expressed as an acceptable range.

What Can Go Wrong

Like other semiconductors, a triac is heat sensitive. Usual precautions should be taken to allow sufficient ventilation and heat sinking, especially when components are moved from an open prototyping board to an enclosure in which crowding is likely.

Unexpected Triggering Caused by Heat

On a datasheet, a value for triggering current is valid only within a recommended temperature range. A buildup of heat can provoke unexpected triggering.

Low-Temperature Effects

Significantly higher gate current will be required by a triac operating at low temperatures. It is quite possible that the component will need twice as much current at 25° C compared with 100° C, junction temperature. If the triac receives insufficient current, it will not turn on.

Wrong Type of Load

If an **incandescent lamp** is replaced with a **fluorescent light** or **LED area lighting**, a pre-existing triac may no longer work as a dimmer. Fluorescent lamps will have some inductance, and may also provide a capacitive load, either of which will interfere with the normal behavior of a triac.

The light output of an LED varies very differently compared with the light output of an incandescent bulb, in response to reduction in power. Therefore an LED should be dimmed using pulse-width modulation that is appropriate for its output characteristics. A triac is generally not suitable.

Wrongly Identified Terminals

A triac is often thought of as a symmetrical device, because it is designed to switch AC current using either positive or negative voltage at the gate. In reality, its behavior is asymmetrical, and if it is installed "the wrong way around" it may function erratically or not at all.

Failure to Switch Off

As already noted (see "Charge Storage" on page 23), a triac will tend to suffer from *charge storage* if there is insufficient time between the end of one half-cycle and the beginning of the next. A component that works with a resistive load may cease to function if it is used, instead, to power an inductive load.

solid-state relay

4.

A **solid-state relay** is less-commonly referred to by its acronym, *SSR*. It is sometimes regarded as an **optocoupler**, but in this encyclopedia the two components have separate entries. An optocoupler is a relatively simple device consisting of a light source (usually an **LED**) and a light sensor, in one package. It is used primarily for isolation rather than to switch a high current. A solid-state relay can be thought of as a substitute for an electromagnetic **relay**, usually has additional components in its package, and is intended to switch currents of at least 1A.

A component that works like a solid-state relay but only switches a 5V (or lower) logic signal may be referred to as a *switch*, even though it is entirely solid-state. This type of component is included in this entry because it functions so similarly to a solid-state relay.

OTHER RELATED COMPONENTS

- **electromagnetic relay** (see Volume 1)
- **optocoupler** (see Chapter 5)

What It Does

A **solid-state relay** (SSR) is a semiconductor package that emulates an electromagnetic **relay** (see Volume 1). It switches power on or off between its output terminals in response to a smaller current and voltage between its input terminals. Variants can switch AC or DC and may be controlled by AC or DC. An SSR functions as a *SPST switch*, and is available in normally open or normally closed versions. SSRs that function as an SPDT switch are relatively unusual and actually contain more than one SSR.

No single schematic symbol has been adopted to represent a solid-state relay, but some alternatives are shown in Figure 4-1:

Top
>An unusually detailed depiction of an SSR that switches DC current using MOSFETS. Symbols for this device often omit the diodes

on the output side and may simplify the MOSFET symbols.

Bottom left
>An SSR that uses an internal **triac** to switch AC. The box labeled 0x indicates that this is a *zero-crossing* relay, meaning that it switches when alternating voltage crosses the 0V level from positive to negative or negative to positive.

Bottom right
>A generic SSR, showing a symbol for a normally open relay, although whether it is designed for AC or DC is unclear.

Advantages

- Great reliability and long life.
- No physical contacts that are vulnerable to arcing and erosion or (under extreme conditions) that could weld themselves together.

- Very fast response, typically 1μs on and 0.5μs off.

- Very low power consumption on the input side, as low as 5mA at 5VDC. Many solid-state relays can be driven directly from logic chips.

- Lack of mechanical noise.

- No contact bounce; a clean output signal.

- No coil that would introduce back EMF into the circuit.

- Safe with flammable vapors, as there is no sparking of contacts.

- Often smaller than a comparable electromagnetic relay.

- Insensitive to vibration.

- Safer for switching high voltages, as there is complete internal separation between input and output.

- Some variants work with input control voltages as low as 1.5VDC. Electromagnetic relays typically require at least 3VDC (or more, where larger relays are required to switch higher currents).

Disadvantages

- Less efficient; its internal impedance introduces a fixed-value voltage drop on the output side (although this may be negligible when switching higher voltages).

- Generates waste heat in its "on" mode, in accordance with the voltage drop.

- Passes some *leakage* current (usually measured in microamps) on the output side when the relay is supposed to be "off."

- A DC solid-state relay usually requires observation of polarity on the output side. An electromagnetic relay does not.

- Brief voltage spikes on the input side, which would be ignored by a slower electromagnetic relay, may trigger a solid-state relay.

- More vulnerable than an electromagnetic relay to surges and spikes in the current that is switched on the output side.

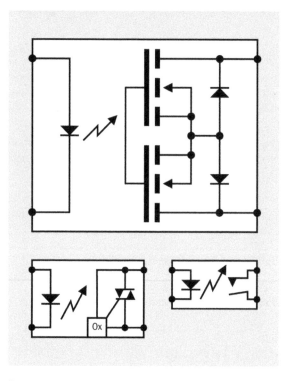

Figure 4-1. *Schematic symbols for solid-state relays have not been standardized. See text for details.*

How It Works

Almost all modern SSRs contain an internal **LED** (light-emitting diode, see Chapter 22) which is switched on by the control input. Infrared light from the LED is detected by a sensor consisting of one or more **phototransistors** or **photodiodes**. In a relay that controls DC current, the sensor usually switches a **MOSFET** (see Volume 1) or an **SCR** (silicon-controlled rectifier—see Chapter 1). In relays that control AC current, a **triac** (see Chapter 3) controls the output. Because the input side and the output side of the SSR are linked only by a light signal, they are electrically isolated from each other.

The MOSFETs require so little power, it can be provided entirely by light falling on an array of 20 or more photodiodes inside the SSR package.

Typical solid-state relays are shown in Figures 4-2 and 4-3.

Figure 4-2. *A solid-state relay capable of switching up to 7A DC. See text for a detailed description.*

The Crydom DC60S7 accepts a control voltage ranging from 3.5VDC to 32VDC, with a typical input current of less than 3mA. Maximum turn-on time is 0.1ms and maximum turn-off time is 0.3ms. This relay can switch up to 7A and tolerates a surge of up to twice that current. It imposes a voltage drop of as much as 1.7VDC, which can become a drawback when switching voltages that are significantly lower than its maximum 60VDC. The electronics are sealed in thermally conductive epoxy, mounted on a metal plate approximately 1/8" thick which can be screwed down onto an additional heat sink.

The Crydom CMX60D10 tolerates a more limited range of control voltages (3VDC to 10VDC) and requires a higher input current of 15mA at 5VDC. However, its very low maximum on-state resistance of 0.018Ω imposes a much smaller voltage drop of less than 0.2 volts when passing 10A. This results in less waste heat and enables a *single-inline package* (SIP) without a heat sink. The

CMX60D10 weighs 0.4 ounces, as opposed to the 3 ounces of the DC60S7. Relays from other manufacturers use similar packaging and have similar specifications.

Figure 4-3. *A solid-state relay capable of switching up to 10A. Its lower internal resistance results in less waste heat and enables a smaller package. See text for a detailed description.*

Variants

Many solid-state relays have protective components built into the package, such as a varistor on the output side to absorb transients. Check datasheets carefully to determine how much protection from external components may be necessary when switching an inductive load.

Instantaneous versus Zero Crossing

A *zero crossing* SSR is one that (a) switches AC current and (b) will not switch "on" until the instant when the AC voltage crosses through 0V. The advantages of this type are that it does not have to be built to switch such a high current, and creates minimal voltage spike when the switching occurs.

All SSRs that are designed to switch AC will wait for the next voltage zero crossing before switching to their "off" state.

NC and NO Modes

Solid-state relays are SPST devices, but different models may have a normally closed or normally open output. If you require double-throw operation, two relays can be combined, one normally closed, the other normally open. See Figure 4-4. A few manufacturers combine a normally closed relay and a normally open relay in one package, to emulate a SPDT relay.

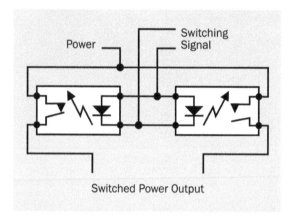

Figure 4-4. *A normally closed solid-state relay can be paired with a normally open solid-state relay to emulate a SPDT switch. This combination is available in a single package from some manufacturers.*

Packaging

High-current solid-state relays are often packaged with screw terminals and a metal base that is appropriate for mating with a *heat sink*. Some are sold with heat sinks integrated. Spade terminals and crimp terminals may be optional. The Crydom DC60S7 shown in Figure 4-2 is an example. This type of package may be referred to as *industrial mount*.

Lower-current solid-state relays (5A or less), and those with a very low output resistance, may be packaged with single-inline pins for through-hole mounting in circuit boards.

Solid-State Analog Switch

DIP packaging may be used for solid-state relays that are designed for compatibility with the low voltages and currents of logic chips. This type of

component may be referred to simply as a *switch*. The 74HC4316 is an example, pictured in Figure 4-5.

Figure 4-5. *This DIP package contains four "switches" that function as solid-state relays but are restricted to low voltages and currents, compatible with logic chips. See text for details.*

Typically the control voltage and the switched voltage are limited between +7V and −7V, with a maximum output current of 25mA. Each internal switch has its own Control pin, while an additional Enable pin forces all switches into an "off" state if its logic state is high. The simplified functionality of this component is illustrated in Figure 4-6, without showing internal optical isolation.

The "on" resistance of each internal pathway will be approximately 200Ω when the component is powered with +5VDC on the positive side and 0VDC on the negative side. This resistance drops to 100Ω if the negative power supply is -5VDC.

If all of the outputs from the chip are shorted together, it functions as a **multiplexer** (see Chapter 16). In fact, this type of switch component is often listed in catalogs as a multiplexer, even though it has other applications.

Because the component tolerates equal and opposite input voltages, it is capable of switching AC.

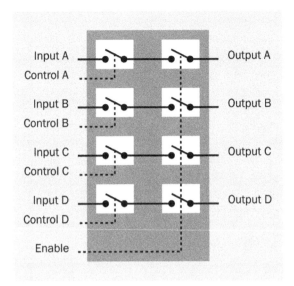

Figure 4-6. *The functionality of a chip containing four solid-state analog switches. A high state on a Control pin closes its associated switch. The Enable pin must be held low for normal operation; a high Enable state forces all the switches into the "off" position. If the outputs are tied together, this component can function as a multiplexer.*

Values

Industrial-mount solid-state relays typically can switch currents ranging from 5A to 500A, with 50A being very common. The higher-current relays mostly require DC control voltage; 4V to 32V are typical, although some versions can go much higher. They contain an SCR or triac to switch AC.

Smaller solid-state relays in SIP, DIP, or surface-mount packages often use MOSFETs on the output side, and are often capable of switching up to 2A or 3A. Some can switch either AC or DC, depending on the way the output is wired. The LED on the input side may require as little as 3mA to 5mA for triggering.

How to Use It

Solid-state relays find their primary uses in telecommunications equipment, industrial control systems and signalling, and security systems.

The component is very simple externally. Power on the input side can come from any source capable of delivering the voltage and current specified by the manufacturer, and any device that doesn't exceed maximum current rating can be connected to the output side, so long as provision is made for suppressing back-EMF from an inductive load, as shown in Figure 4-7. Often a solid-state relay can be substituted directly for an electromagnetic relay, without modifying the circuit.

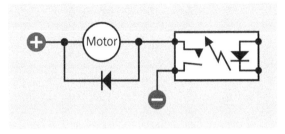

Figure 4-7. *Use of a diode around an inductive load, to protect a solid-state relay from back-EMF.*

Solid-state relays are heat sensitive, and their rating for switching current will diminish as their temperature increases. Manufacturer datasheets will provide specific guidance. Using a heat sink will greatly improve the performance. Bear in mind that the relay generates heat continuously while it is in its "on" mode—about 1 watt per ampere.

Because it requires so little current on the input side (typically no more than 15mA), a solid-state relay can usually be driven directly by chips such as *microcontrollers* that would not be able to activate an equivalent electromagnetic relay.

Applications may take advantage of the solid-state relay's reliability, immunity to vibration, lack of contact sparking, freedom from coil-induced surges on the input side, and lack of contact bounce on the output side. A solid-state relay is ideal within digital equipment that is sensitive to power spikes. It may switch a fuel pump that handles volatile, flammable liquids, or a wastewater pump in a basement subject to flooding (where long-term zero-maintenance reliability is necessary, and contact corrosion could be a risk in electromagnetic relays). Small solid-state relays can switch motors in robots or appliances where vibration is common, and are often used in arcade games.

What Can Go Wrong

Overheating Caused by Overloading

Relays must be *derated* when used at operating temperatures above the typical 20 or 25° C for which their specification applies. In other words, the sustained operating current must be reduced, usually by an amount such as 20% to 30% for each 10-degree increase in ambient temperature. Failure to observe this rule may result in failure of the component. Burnout may also occur if a high-current solid-state relay is used without a heat sink, or the heat sink isn't big enough, or *thermal compound* is not applied between the solid-state relay and the heat sink.

Overheating Caused by Bad Terminal Contact

If the screw terminals on the output side of a high-current solid-state relay are not tightened sufficiently, or if there is a loose spade terminal, or if a crimped connection isn't crimped tightly enough, the poor contact will create electrical resistance, and at high currents, the resistance will create heat, which can cause the solid-state relay to overheat and burn out.

Overheating Caused by Changing Duty Cycle

If a high-current solid-state relay is chosen for an application where it is in its "on" state only half the time, but the application changes during product development so that the solid-state relay is in its "on" state almost all the time, it will have to dissipate almost twice as much heat. Any time the duty cycle is changed, heat should be considered. The possibility of the relay being

used in an unconventional or unexpected manner should also be considered.

Overheating Caused by Component Crowding

Overheating increases dramatically when components are tightly crowded. At least 2cm (3/4") should be allowed between components.

Overheating in Dual Packaging

When a package contains two solid-state relays, the additive effects of the heat created by each of them must be considered.

Reverse-Voltage Burnout

Because a solid-state relay is more sensitive to *back-EMF* than an electromagnetic relay, greater care should be used to protect it from reverse voltage when switching inductive loads. A *protection diode* should be used, and a *snubber* can be added between its output terminals, if it is not included inside the relay package.

Low Voltage Output Current May Not Work

Unlike electromagnetic relays, solid-state relays require some voltage on the output side to enable their internal operation. If there is no voltage, or only a very low voltage, the SSR may not respond to an input. The minimum voltage required on the output side is usually specified in a datasheet.

To test a solid-state relay, apply actual voltages on input and output sides and use a load such as an incandescent light bulb. Merely applying a meter on the output side, set to measure continuity, may not provide sufficient voltage to enable the relay to function, creating the erroneous impression that it has failed.

Inability to Measure AC Output

When a multimeter is used to test continuity across the output of an AC-switching solid-state relay of zero-crossing specification, the meter will generate enough voltage to prevent the solid-state relay from finding zero voltage across its output terminals, and consequently the solid-state relay won't switch its output.

Relay Turns On but Won't Turn Off

When a solid-state relay controls a relatively high-impedance load such as a small **solenoid** (see Volume 1) or a **neon bulb** (see Chapter 19), the relay may switch the device on but will seem unable to switch it off. This is because the leakage current of the solid-state relay, in its "off" state, may be just enough to maintain the load in its "on" state.

If an SSR containing a triac is used erroneously to switch DC, it will not be able to switch off the current.

Relays in Parallel Won't Work

Two solid-state relays usually cannot be used in parallel to switch twice as much current. Because of small manufacturing variances, one relay will switch on a moment before the other. When the first relay is on, it will divert the load current away from the second relay. The second relay needs a small amount of current on its output side, to function. Without any current, it will not switch on. This means the first relay will pass the total current without any help from the second relay, and will probably burn out, while the second relay does nothing.

Output Device Doesn't Run at Full Power

A solid-state relay imposes a voltage reduction on its output side. This will be a fixed amount, not a percentage. When switching 110V, this difference may be negligible; when switching 12V, it may deliver only 10.5V, which represents enough of a drop to cause a motor or a pump to run noticeably more slowly. The internal switching device inside the relay (MOSFET, triac, SSR, or bipolar transistor) will largely determine the voltage drop. Check the manufacturer's datasheet before using the relay.

Solid-State Relays and Safety Disconnects

Because a solid-state relay always allows some leakage in its "off" state, it can still deliver a shock when used to switch high voltages. For this reason, it may not be suitable in a safety disconnect.

optocoupler

5

Sometimes known as an *optoelectronic coupler*, *opto-isolator*, *photocoupler*, or *optical isolator*.

A **solid-state relay** is sometimes referred to as an **optocoupler**, but in this encyclopedia it has a separate entry. An optocoupler is a relatively simple device consisting of a light source (usually an LED) and a light sensor, both embedded in one package. It is used primarily for isolation rather than to switch a high current. A solid-state relay can be thought of as a substitute for an electromagnetic **relay**, usually has additional components in its package, and is intended to switch currents of at least 1A.

OTHER RELATED COMPONENTS

- **electromagnetic relay** (see Volume 1)
- **solid-state relay** (see Chapter 4)

What It Does

An optocoupler allows one section of a circuit to be electrically isolated from another. It protects sensitive components, such as logic chips or a microcontroller, from voltage spikes or incompatible voltages in other sections of a circuit. Optocouplers are also used in medical devices where a patient has to be protected from any risk of electric shock, and are used in devices which conform with the MIDI standard for digital control of music components.

In Figure 5-1, three possible applications for an optocoupler are suggested:

Top
> The output from a logic chip passes through an optocoupler to an inductive load such as a relay coil, which may create voltage spikes that would be hazardous to the chip.

Center
> The noisy signal from an electromagnetic switch passes through an optocoupler to the input of a logic chip.

Bottom
> The low-voltage output from a sensing device on a human patient passes through an optocoupler to some medical equipment, such as an EEG machine, where higher voltages are used.

Internally, an optocoupler works on the same principle as a **solid-state relay**. An LED is embedded on the input side, shining light through an interior channel or transparent window to a sensing component that is embedded on the output side. Because the only internal connection is a light beam, the input and output of the optocoupler are isolated from each other.

33

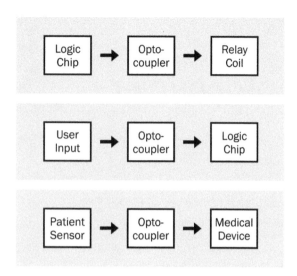

Figure 5-1. *Possible applications for a photocoupler. See text for details.*

Isolation transformers were traditionally used for this purpose prior to the 1970s, when optocouplers became competitive. In addition to being smaller and cheaper, an optocoupler can also pass slow-changing signals or on-off DC states which a transformer would ignore.

More recently, inductive and capacitive coupling components have become available in surface-mount packages that are competitive with optocouplers for high-speed data transfer. They also claim to be more durable. Because of the gradual reduction in output from an LED, the performance of an optocoupler degrades over time, and is typically rated for up to 10 years.

How It Works

The LED in an optocoupler almost always emits light in the near-infrared part of the spectrum, and is matched to the sensitivity of a **phototransistor**, or a **photodiode**, or (less often) a **photoresistor** that provides the output. Photosensitive **triacs** and **SCRs** are also sometimes used.

The most common type of optocoupler uses a bipolar phototransistor with an open-collector output. Schematic symbols for this type are shown in Figure 5-2:

Top left
> The most common generic form.

Top right
> Two diodes on the input side allow the use of alternating current.

Center left
> An additional terminal allows addition of bias to the photosensitive base of the output transistor, to reduce its sensitivity.

Center right
> An Enable signal can be used as the input to the NAND, suppressing or enabling the output.

Bottom left
> A *photodarlington* allows higher emitter current.

Bottom right
> Relatively uncommon, and is also used for a **solid-state relay**.

In each symbol, the diode is an LED, and the zig-zag arrow indicates light that is emitted from it. A pair of straight arrows, or wavy arrows, may alternatively be used.

An optocoupler in through-hole DIP format is shown in Figure 5-3.

An *optical switch* can be thought of as a form of optocoupler, as it contains an LED opposite a sensor. However, the LED and the sensor are separated by an open slot, to allow a thin moving object to pass through, interrupting the light beam as a means of detecting the event. It is categorized as a *sensor* in this encyclopedia, and will be found in Volume 3.

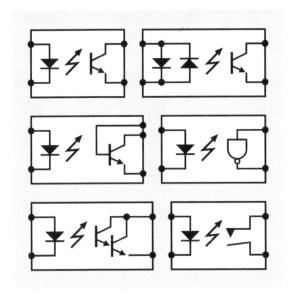

Figure 5-2. *Six variants of schematic symbols that may be used to represent an optocoupler. See text for details.*

Figure 5-3. *An optocoupler in through-hole 8-pin DIP format.*

Variants

Internal Sensors

Historically, a photoresistor (often referred to as a *photocell*) was the first type of sensor to be used. It has a more linear response than other sensor types, but its response is much slower. It is still found in audio applications. "Stomp box" pedals used by guitarists typically contain an optocoupler that employs a photoresistor, and are valued for their linearity and their immunity from the mechanical wear, contamination, and "scratchiness" that builds up over time in a **potentiometer**. Optocouplers also eliminate *ground loops*, which tend to be induced by small differences in ground potential, introducing hum or buzz in audio applications when two or more power supplies are tied together.

The type of optocoupler that contains a photoresistor and is commonly used by musicians was initially trademarked as a *Vactrol*, and that term is still used generically. Vactrols have also been used to provide audio compression in telephone voice networks, and were used in photocopiers and photographic exposure meters, but these applications are now obsolete.

Photoresistors are becoming uncommon because of their cadmium content, which is unlawful in many countries (especially in Europe) because of its environmental toxicity.

A photodiode provides the fastest response time in an optocoupler, limited primarily by the characteristics of the LED that shines light upon it. A *PIN diode* can respond in less than a nanosecond; its acronym is derived from its fabrication from p-type and n-type semiconductor layers with an *intrinsic* layer connecting them. This additional layer can be responsive to light. When the diode is slightly reverse-biased, a photon entering the intrinsic layer can dislodge an electron, enabling current to flow. The reverse bias enlarges the active area and enhances the effect. In this mode, the PIN acts like a photoresistor, appearing to reduce its resistance in response to light.

When the PIN is used in *photovoltaic* mode, no bias is applied, and the component actually generates a small voltage (less than 1VDC), like a solar cell, in response to incoming light. Where an optocoupler uses a MOSFET on its output side, as many as 30 photodiodes may be connected in series to develop the necessary threshold voltage to trigger the transistor. This arrangement is common in **solid-state relays**.

A bipolar phototransistor is a slower-speed device but is still usually capable of a 5µs response time or better. Its open collector requires external voltage and a *pull-up resistor* to deliver a positive output so long as the phototransistor is nonconductive. When the LED turns on, the phototransistor sinks current, effectively creating a low output. In this way, the optocoupler functions like an *inverter*, although some variants include a noninverting output.

Basic Optocoupler Types

An optocoupler with *high linearity* will respond more proportionally to variations in current to its LED. *High Speed* optocouplers are used for high-frequency data transfer. *Logic-output* optocouplers have a clean high/low output transition, rather than an *analog output*, which varies with fluctuations in the input. Linearity is of importance only where an optocoupler is being used to transmit an analog signal with some fidelity. Some logic-output optocouplers provide the function of a Schmitt trigger on their output side.

While optocouplers are available in various package formats, the DIP style with six or eight pins remains popular, providing sufficient physical space for the LED, the sensor, and a light channel, while providing good electrical isolation.

Variants may have two or four optocouplers combined in one package. A *bidirectional* optocoupler may consist of two optocouplers in parallel, inverted with respect to each other.

Values

In a datasheet, the characteristics of primary importance in an optocoupler are:

- CTR is the Current Transfer Ratio, the ratio of maximum output current to input current, expressed as a percentage. With a bipolar phototransistor output, 20% is a typical minimum CTR. With a photodarlington output,

the CTR may be 1,000% but the bandwidth is much lower—the response time may be measured in microseconds rather than nanoseconds. Optocouplers with a photo-diode output have a very low CTR, and their output is in microamps. However, they provide the most linear response.

- $V_{CE(MAX)}$ is the maximum collector-emitter voltage difference (in an optocoupler with a bipolar phototransistor output). Values from 20 to 80 volts are common.

- V_{ISO} is the maximum potential difference, in VDC, between the two sides of the optocoupler.

- I_{MAX} is the maximum current the transistor can handle, generally in mA.

- Bandwidth is the maximum transmittable signal frequency, often in the range of 20kHz to 500kHz.

The LED in an optocoupler typically requires 5mA at a forward voltage of 1.5V to 1.6V.

The maximum collector current on the output side of an optocoupler is unlikely to be higher than 200mA. For higher output currents, a solid-state relay should be considered. It provides photo-isolation on the same basis as an optocoupler, but high-current versions tend to be considerably more expensive.

How to Use It

The primary purpose of an optocoupler is to provide protection against excessive voltage—from transients, incompatible power supplies, or equipment with unknown characteristics. If a device is designed to be plugged into a USB port on a computer, for instance, the computer may be isolated via an optocoupler.

A series resistor for the LED is not built into most optocouplers, because the value of the resistor will depend on the input voltage that is used. Care must be taken to determine what the maximum voltage on the input side will be, and a series resistor should be chosen to reduce current appropriately. Allowance should be made for some degradation in the performance of the LED over time.

For an optocoupler with an open-collector output, a pull-up resistor is necessary in most applications. The voltage from the optocoupler must be matched to the input requirements of other components, and the collector current must remain within the specified limits. Some trial and error in resistor selection may be necessary.

In Figure 5-4, a schematic shows typical component values in a test circuit using a pushbutton as input. The separation of the two power supplies is emphasized by the different color shades used for the positive and negative symbols. Although the input side and the output side of an optocoupler may be used with a common ground, this defeats its purpose in providing complete isolation between the sections of the circuit.

The pinouts for an optocoupler must be checked carefully in the manufacturer's datasheet. While the input for an 8-pin DIP chip is usually applied to pins 2 and 3, the output pin functions are not standardized and will vary depending on the internal configuration of the chip. An optocoupler such as the Optek D804, with an enable function using an internal NAND gate, requires its own power supply.

Where an optocoupler allows an external connection to the base of its internal bipolar output phototransistor, reverse bias applied to this pin will decrease the sensitivity of the optocoupler but can increase its immunity to noise on the input side.

What Can Go Wrong

Overload conditions on the input or the output side of an optocoupler will be the most likely cause of failure.

Age

Because optocouplers are typically rated for only 10 years of average use, the age of a component may cause it to fail.

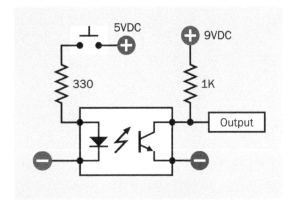

Figure 5-4. *Typical values for a series resistor (to protect the LED) and pull-up resistor (to control current and voltage on the output side) in an optocoupler test circuit.*

LED Burnout

Because the LED is hidden inside the component, there is no immediate indication of its performance. A meter can be inserted into the circuit on the input side to determine if current is passing through the LED. A meter set to measure volts can be used to discover whether the LED is imposing a normal voltage drop. While significant overload will cause immediate burnout, slightly exceeding the current rating of the LED may have more pernicious consequences, as the LED may not fail until days or weeks have passed without any sign of trouble. The failure of the optocoupler will be unexpected and difficult to determine.

Transistor Burnout

Here again the damage caused by excessive current may be progressive, occurring over a prolonged period. The easiest way to test an optocoupler that may have failed is by removing it from the circuit. A socketed DIP package is preferable for this purpose.

comparator

Although a **comparator** has the same schematic symbol as an **op-amp**, their applications differ and they are described in separate sections of this encyclopedia.

This entry describes an analog comparator. A *digital comparator* is very different, being a logic chip that compares two binary numbers that can be referred to as A and B. Outputs from the chip indicate whether A>B or A<B or A=B. The digital comparator does not have an entry in this encyclopedia.

OTHER RELATED COMPONENTS

- **op-amp** (see Chapter 7)

What It Does

A **comparator** is an integrated circuit chip that compares a variable voltage on one input pin with a fixed, reference voltage on a second input pin. Depending which voltage is higher, the output from the comparator will be high or low.

The output will make a clean transition between two fixed values, even if the input is infinitely variable. Thus the comparator can function as an *analog-digital converter*, as shown in Figure 6-1.

Because the output voltage range can be adjusted up or down independently of the input range, a comparator can also function as a *voltage converter*.

Hysteresis

If positive feedback is added through external resistors, *hysteresis* can be introduced. We may imagine a *hysteresis zone* extending above and below the reference voltage level. Small input variations that occur within the zone will be ignored. The comparator only reacts when the input signal emerges above or below the hysteresis zone. When the input signal returns into the hys-teresis zone, this event also will be ignored. The concept is illustrated graphically in Figure 6-2. A circuit to create hysteresis is shown in Figure 6-10.

How It Works

The schematic symbol for a comparator is shown in Figure 6-3. This seems identical to the symbol for an op-amp, described in Chapter 7, but an op-amp is traditionally a *dual-voltage* device using positive and negative power sources that are equal and opposite, in addition to a zero value midway between the two. Modern comparators mostly use a conventional single voltage, and therefore the negative symbol used in comparator schematics throughout this section of the encyclopedia represents 0 volts. It has the same meaning as the ground symbol found in many schematics elsewhere.

The two inputs to a comparator are described as *inverting* and *noninverting* (for reasons explained later). Confusingly, these are identified with plus and minus symbols inside the triangle that represents the component. These plain black-and-white symbols have nothing to do with the power supply.

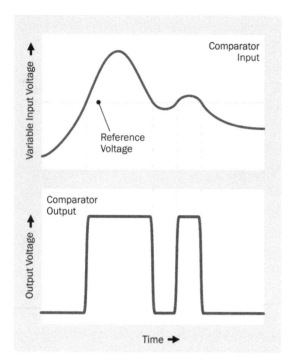

Figure 6-1. *The basic behavior of a simple comparator is shown here.*

Often, in schematics, the power supply is not shown, because it is assumed to be present. However, all comparators require a power supply in order to function.

The basic internal and external connections used in conjunction with a typical comparator are shown in Figure 6-4.

The potentiometer at top left is often a *trimmer*, to fine-tune a reference voltage. The variable input can come from a sensor or any other device capable of delivering a voltage up to the limit set by V1.

The output is often an *open collector* from an internal bipolar transistor, as shown in the figure.

Note that as many as three different voltages can be used, as indicated by the different colors associated with V1, V2, and V3. However, they must share a common ground to enable the comparator to make valid comparisons.

When the noninverting input exceeds the voltage of the inverting input, the output transistor goes into its "off" state, and blocks current from an external *pullup resistor*. Because the current from the resistor now has nowhere else to go, it is available to drive other devices attached to the comparator output, and the output appears to be high.

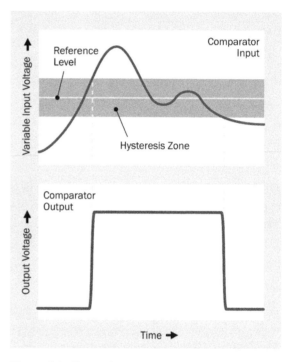

Figure 6-2. *The performance of a comparator shown in the previous figure can be modified by the addition of hysteresis. Small variations that occur within the hysteresis zone are ignored.*

When the noninverting input falls below the voltage of the inverting input, the transistor becomes conductive, and sinks almost all the current from the pullup resistor, assuming other devices attached to the output have a relatively high impedance. The output from the comparator now appears to be low.

This can be summed up as follows:

- When a variable voltage is applied to the noninverting input, and it rises *above* the reference voltage applied to the *inverting input*, the output transistor turns off, and the comparator delivers a *high* output.

- When a variable voltage is applied to the noninverting input, and it falls *below* the reference voltage applied to the *inverting input*, the output transistor turns on, and the comparator delivers a *low* output.

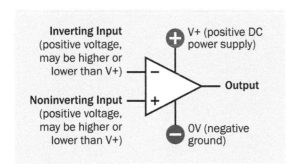

Figure 6-3. *The symbol for a comparator is the same as the symbol for an op-amp, even though they often require different types of power supply and their functions are significantly different.*

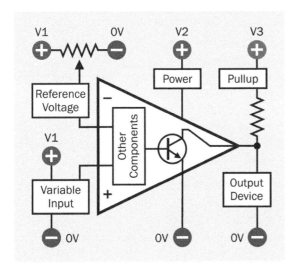

Figure 6-4. *Connections to a comparator, and their functions.*

If the reference voltage and the variable voltage are swapped between the input pins, the behavior of the comparator is reversed. This relationship is illustrated in Figure 6-5. When a voltage transition is applied to the *inverting* input, the transition is *inverted* at the output.

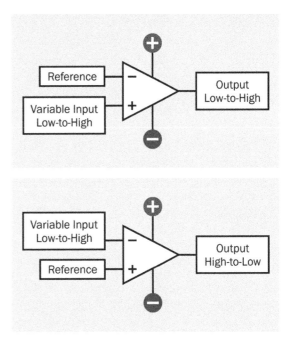

Figure 6-5. *Depending which input pin is used for the reference voltage, and which input carries a variable voltage, the comparator output either follows the variable voltage or inverts it.*

Placement of the plus and minus signs inside the comparator symbol may vary. Most often, the minus sign is above the plus sign, as shown in all the schematics here. Sometimes, however, for convenience in drawing a schematic, the plus sign may be shown above the minus sign. Regardless of their placement, the plus sign always identifies the noninverting input, and the minus sign always identifies the inverting input. To avoid misinterpretations, schematics should be inspected carefully.

Where a power supply for the comparator is shown, the positive side is always attached to the upper edge of the symbol, while 0V ground is always attached to the lower edge.

Differences from an Op-Amp

Saturation versus linearity

The output of a comparator is optimized for saturation (high or low, without intermediate levels, using positive feedback). The output of an op-amp is optimized for linearity (faithful reproduction of nuances in the input, using negative feedback).

Output mode

The majority of comparators have open-collector outputs (or open-drain outputs in CMOS devices) where the voltage is established by a pullup resistor. This can be adjusted for compatibility with other components, especially 5VDC logic. Only a minority have push-pull amplifier outputs that require no pullup resistor. By comparison, among op-amps, a push-pull output that functions as a voltage source is the traditional default.

Faster response

A comparator responds more quickly than an op-amp to changes in input voltage, if the op-amp is used in the role of a comparator. The comparator is primarily a switching device, not an amplifier.

Hysteresis

This is generally desirable in a comparator, for reasons already explained, and some components are designed with hysteresis built in. This feature is undesirable in an op-amp, as it degrades sensitivity.

Open-loop operation

(i.e., without feedback) this can be used with a comparator. An op-amp is intended for use in closed-loop circuits (i.e., with feedback), and manufacturers will not specify its performance in an open loop.

As previously noted, a comparator usually requires a single-voltage power supply, while an op-amp often requires a dual-voltage power supply.

Variants

Where a comparator uses a MOSFET output transistor, it may have an open-drain output, which requires a pullup resistor, as with an open-collector output.

Some comparators have a *push-pull* output, capable of supplying output current (usually a small amount). In these instances, no pullup resistor is necessary or desirable. The output voltage range will be closest to *rail-to-rail* values (i.e., the range of the power supply) where MOSFETs are used for the output, as MOSFETs impose a smaller voltage drop than bipolar transistors.

The advantage of an open collector (or open drain) relative to a push-pull output is that it allows the output voltage to be set independently of the power supply voltage. Another advantage is that multiple outputs can be connected in parallel, as in a *window comparator* circuit (described below).

Some comparators incorporate a reference voltage on the chip, based on the power supply to the chip. In this case, a separate reference voltage does not have to be supplied, and the component will draw less current.

Many chips are available containing two or more comparators. This is often expressed as the number of *channels* in the component. A *dual comparator* typically allows two different voltage sources for the outputs of the comparators. They will share the same 0V ground, however. Chips such as the LM139 and LM339 contain four comparators, and are available in through-hole or surface-mount formats. They have become a generic choice, costing less than $1 apiece.

An LM339 comparator chip is shown in Figure 6-6. This is a quad chip, meaning that it contains four comparators. They share a common power supply. The chip is TTL and CMOS compatible, is typically powered by 5VDC, but can be driven by up to 36VDC. The input differential voltage range also extends up to 36V.

Figure 6-6. *The LM339 quad comparator chip, shown here, was introduced long ago but remains widely used.*

Some comparators have an internal *latch* function that is accessed by a dedicated pin. The latch-enable signal forces the comparator to assess its inputs and hold an appropriate output which can then be checked by other components.

Values

In a datasheet, V_{IO} (also referred to as V_{OS}) is the *input offset voltage*. This is a small voltage, in addition to the reference voltage, which the comparator will require to *toggle* its output in either direction, up or down. Figure 6-7 shows this graphically. V_{IO} sets the limit of *resolution* of the comparator, which will not respond unless the input voltage exceeds the reference voltage by this amount. A smaller value for V_{IO} is better than a larger value. Common values for V_{IO} range from 1mV to 15mV. The actual offset voltage tends to vary between one sample of a component and another. V_{IO} is the maximum allowed value for a component.

Because the comparator will not respond until the reference voltage is exceeded by V_{IO}, the output pulse width will be narrower than if the comparator reacted at the point where the variable voltage input was precisely the same as the reference voltage.

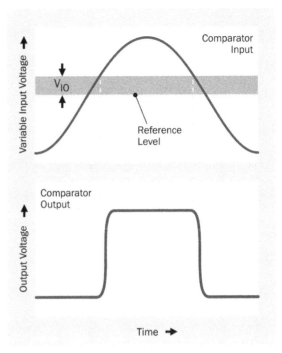

Figure 6-7. *The input offset voltage is the very small voltage that a comparator requires, additional to the reference input voltage, before it will toggle its output from low to high or high to low.*

V_{TRIP+} and V_{TRIP-} are the *rising* and *falling* voltages, respectively, that will trip the comparator output where the comparator exhibits some innate hysteresis without an external feedback loop. They are also referred to as *Lower State Transition Voltage* (LSTV) and *Upper State Transition Voltage* (USTV).

V_{HYST} is the *hysteresis range* defined as V_{TRIP+} minus V_{TRIP-}. The relationship is shown graphically in Figure 6-8.

A_{VD} is the *voltage gain* of a comparator, in which the letter "A" can be thought of as meaning "amplification." The gain is measured as a maximum ratio of output voltage to input voltage. Typically it ranges from 40 to 200.

Supply voltage for modern comparators is often low, as the components are used in surface-mount format for battery-powered devices where low power consumption is a primary concern. Thus, 3VDC is common as a power requirement, and 1.5VDC comparators are available. Still, older chips can use as much as 35VDC.

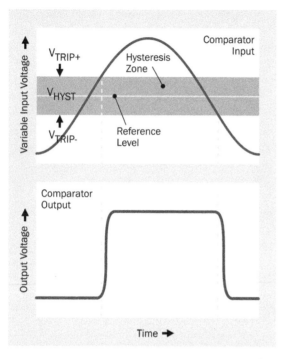

Figure 6-8. *The value of V$_{TRIP}$ shows the hysteresis in a comparator—the range of input voltages, relative to the reference voltage, in which it will not respond.*

Supply current can range from 7mA down to below 1μA.

I$_{SINK}$ is the recommended typical or maximum *sink current* that the component will tolerate, if it has an open-collector output. This value should be considered in relation to the power dissipation, P$_D$.

The *propagation delay* in a comparator is measured from the moment when an input (usually a square wave) reaches the triggering value, to the time when the consequent output reaches 50% of its final value.

When a comparator is driving CMOS logic using a 5VDC power supply, a typical value for a pullup resistor is 100K. It does not have to be lower, because CMOS has such a high input impedance.

How to Use It

In Figure 6-1, a hypothetical comparator responds immediately when the input voltage equals the reference voltage value. However, this is an idealized scenario. A magnified view, in Figure 6-9, suggests that the comparator is likely to respond with *jitter* when the input signal is very close to the reference voltage, because of tiny variations in heat, current, and other variables. This jitter will cause significant problems if the comparator is driving a device such as a relay, directly or indirectly.

Hysteresis eliminates this uncertainty around the transition level of the input, by telling the comparator to ignore small irregularities in the input voltage. Hysteresis is also useful in many situations where larger variations in a sensor input should be ignored. In Figure 6-2, for instance, suppose that the input voltage comes from a temperature sensor. The small bump in the right-hand section of the curve is probably unimportant; it could be caused by someone opening a door, or a person's body heat in brief proximity to the sensor. There's no point in responding to every little event of this type. In this application, the larger, longer-term temperature trend is what matters, and significant hysteresis is appropriate.

Also, if a comparator is being used as a thermostat, to switch a heating system on and off, we do not want the comparator to respond as soon as the temperature rises just a small amount. The heating system should run for a while before it elevates the temperature beyond the hysteresis zone.

The usual way to create hysteresis is with *positive feedback*. In Figure 6-10, a connection from the output of the comparator runs back through a 1M potentiometer to the variable (noninverting) input. The effect that this has is to reinforce the input voltage with the output voltage, as soon as the comparator input goes high. Now the input can diminish slightly without switching off the comparator. But if the input declines significantly, even the feedback from the output voltage won't be sufficient to maintain the variable input at a higher level than the reference voltage. (Remember, the "high" output voltage from the comparator is a fixed value; it does not change in proportion with the input voltage.) Consequently, the output toggles to low. Now the variable input is deprived of help from the comparator output, so it will be low enough that it has to rise considerably to toggle the comparator back on again. During that period, once again, small variations will be ignored.

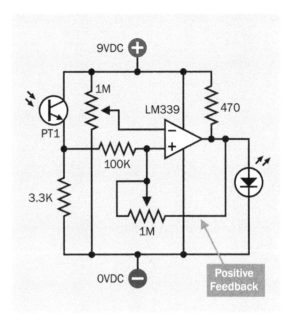

Figure 6-10. *A simple circuit to achieve hysteresis with positive feedback to the variable input of a comparator.*

The 470Ω resistor is the pullup resistor, which protects the LED from excessive current. The lower 1M resistor adjusts the amount of positive feedback, which determines the width of the hysteresis zone.

Values for components may have to be adjusted depending on the supply voltage, the variable input voltage, and other factors. But the principle will remain the same. Note that in the example shown, all the positive voltage sources are identical. In practice, different voltages could be used, so long as they share a common ground.

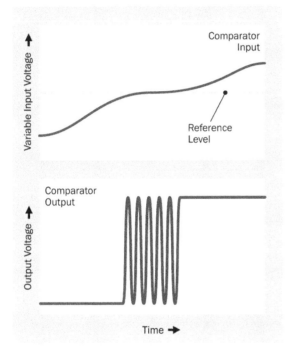

Figure 6-9. *In real-world applications, tiny variations where the variable input voltage crosses the reference voltage can induce jitter in the output from a comparator that has no hysteresis.*

In the schematic, a **phototransistor** (PT1, at left) is in series with a 3.3K resistor to adjust its voltage output to a suitable range. A 1M potentiometer at upper-left is wired as a voltage divider, so that it can establish a reference level that matches the light level that we wish to detect with the phototransistor.

AND gate

A set of open-collector comparators can function jointly as an AND gate, when their outputs are tied together with one pullup resistor. So long as all the output transistors are nonconductive, the output will be high. If just one comparator toggles into conductive mode, the output will be low. See Figure 6-11.

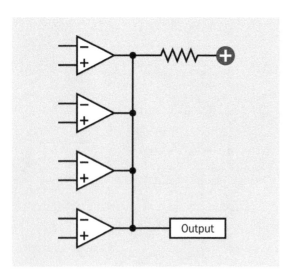

Figure 6-11. *If the outputs of multiple open-collector comparators are tied together with a suitable pullup resistor, they will function as an AND gate.*

Bistable Multivibrator

If positive feedback to the noninverting input of the comparator is sufficiently high, a voltage almost at 0V ground will be required to counter the high output from the comparator—after which, a voltage almost equal to the supply voltage will be needed to turn it back on. In other words, the comparator is behaving like a *bistable multivibrator*, or **flip-flop**.

Relaxation Oscillator

A *relaxation oscillator*, which is a form of *astable multivibrator,* can be created using direct positive feedback in combination with delayed negative feedback. In Figure 6-12, positive feedback goes to the noninverting input, as before, but negative feedback also passes through a 220K resistor to the inverting input of the comparator. A 0.47µF capacitor initially holds the inverting input low, while the capacitor charges. Gradually the capacitor reaches and exceeds the charge on the noninverting input, so the output from the comparator toggles to its low state. This means that its internal transistor is now sinking current, and it discharges the capacitor. Because the noninverting input is being held at a voltage midway between supply and ground by the two 100K re-

sistors forming a voltage divider, eventually the voltage on the inverting input controlled by the capacitor falls below the noninverting voltage, so the cycle begins again.

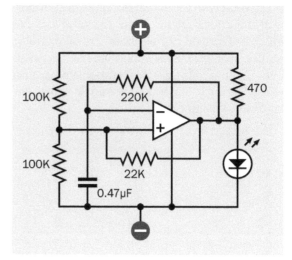

Figure 6-12. *A comparator can be used to create a relaxation oscillator.*

Level Shifter

Where a comparator is used simply to change the level of an input voltage, it can be referred to as a *level shifter*. An example of a level shifter is shown in Figure 6-13, in which a high/low 3VDC logic input is converted to a high/low logic output at 5VDC.

Window Comparator

A *window comparator* is a circuit (not a single component) that will respond to input voltages that deviate outside an acceptable "window" of values. In other words, the circuit responds anytime the variable input is either unacceptably low or unacceptably high.

An example could be an alarm that will sound if a temperature is either too low or too high. In Figure 6-14, two comparators are used to create a window comparator circuit, both sharing a variable voltage input from a sensor. A voltage divider is necessary to establish a higher voltage limit at the noninverting input of the upper comparator, while a separate voltage divider would establish a lower limit at the inverting input of the lower comparator. If an alarm has an appropriate resistance, it can be used instead of a pull-up resistor. The alarm will sound when the output from either comparator is low, which happens if the inverting input has a higher voltage than the non-inverting input.

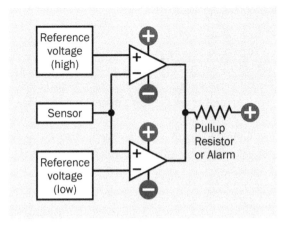

Figure 6-14. *A basic, simplified circuit for a window comparator. See text for details.*

A *continuous converter* changes its output promptly in response to a change in input. This requires continuous current consumption. Because many applications only need to check the output from a comparator at intervals, power can be saved by using a *clocked* or *latched* comparator.

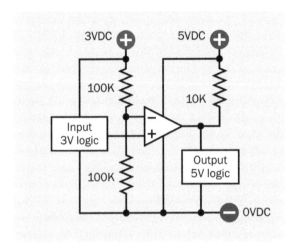

Figure 6-13. *A comparator can be used to convert high and low 3V logic inputs into high and low 5V logic outputs.*

Other Applications

As previously noted, a comparator can be used as a simple *analog-digital converter*. It has "one bit" accuracy (i.e., its output is either high or low).

A comparator can be used as a *zero point finder* when its variable voltage input is attached to an AC signal. The output from the comparator will be toggled whenever the AC signal passes through zero volts. The output will be a square wave (approximately) instead of a sine wave.

What Can Go Wrong

Oscillating Output

The high input impedance of a comparator is vulnerable to stray electromagnetic fields. If the conductors leading to and from the comparator are relatively long, the output can couple capacitively with the input during voltage transitions, causing unwanted oscillations.

The commonly recommended solution to this problem is to add 1µF bypass capacitors to the power supply on either side of the comparator. However, some manufacturers recommend alternatives such as introducing just a small amount of hysteresis, or reducing the value of input resistors to below 10K.

If a chip contains multiple comparators, and one of them is unused, one of its input pins should be tied to the positive side of the voltage supply while the other should be tied to 0V ground, to eliminate the possibility of an oscillating output.

Confused Inputs

A comparator will function if its two inputs are swapped accidentally, but its high and low output states will be the inverse of what is expected. Also, if positive feedback is used, transposed inputs can create oscillations. Because the comparator symbol may appear in a schematic with the noninverting input either below or above the inverting input, the inputs are easily transposed by accident.

One way to remember which way the inputs should be connected is to use the mnemonic: "plus, high, positive." The *plus* input creates a *high* output when the input becomes *more positive* than the reference voltage on the other input. The opposite is less intuitively obvious: the *minus* input creates a *high* output when the input becomes *more negative* than the reference voltage on the other input.

Wrong Chip Type

Different comparators offer different outputs: open collector, open drain, and push-pull. While open collector and open drain function similarly, the pullup resistor value is likely to be different in each case. If a push-pull output is mistakenly connected as if it is open collector or open drain, it will not work correctly, if at all. Different types of comparators must be sorted and stored in clearly labeled bins.

Omitted Pullup Resistor

It is relatively easy to forget to include the pullup resistor on an open-collector output. In this case, when the transistor inside the comparator is in its nonconductive state, the output pin will be floating, with an indeterminate voltage that will create confusing or random results.

CMOS Issues

As usual when using CMOS chips, it is bad practice to allow unconnected, floating inputs. This is an issue where a chip contains multiple comparators, some of which are not being used. The solution recommended by some manufacturers is to tie one input of an unused comparator to the supply voltage, and the other input of the same comparator to ground.

Erratic Output

If positive feedback is insufficient, the comparator output may show signs of *jitter*. Conversely, if the positive feedback is excessive, the comparator may get stuck in an on state or an off state. Feedback must be chosen carefully.

Swapped Voltages

A comparator is often capable of controlling an output voltage that is much higher than that of its power supply. Because both voltages are applied to different pins on the same chip, mistakes can be made quite easily. The chip is likely to be damaged if the voltages are swapped accidentally between the relevant pins.

Heat-Dependent Hysteresis

Remember that the voltages at which the comparator turns on and off will vary slightly with the temperature of the component. This *drift* should be tested by running the comparator at higher temperatures.

op-amp 7

Although a comparator has the same schematic symbol as an **op-amp**, their applications differ and they are described in separate sections of this encyclopedia.

The unabbreviated name for an op-amp is an *operational amplifier*, but this term is seldom used.

OTHER RELATED COMPONENTS

- **comparator** (see Chapter 6)

What It Does

An op-amp is an *operational amplifier* consisting of multiple transistors packaged in an integrated circuit chip. It senses the fluctuating voltage difference between two inputs, uses power from an external supply to amplify that difference, and uses *negative feedback* to ensure that the output is an accurate replica of the input. Its amplification can be adjusted by changing the values of two external resistors.

Op-amps were developed originally using vacuum tubes, for use in analog computers, before the era of digital computing. Their implementation in integrated circuits dates from the late 1960s, when chips such as the LM741 were introduced (lower-noise versions of it still being widely used today). Multiple op-amps in a single package were introduced in the 1970s.

An LM741 is shown in Figure 7-1. Inside the 8-pin, DIP package is a single op-amp.

How It Works

In alternating current, voltages deviate above and below a zero value, which is sometimes referred to as the *neutral* value. This occurs in do-

mestic power supplies and in audio signals, to name two very common examples. A *voltage amplifier* multiplies the positive and negative voltage excursions, using an external power source to achieve this. Most op-amps are voltage amplifiers.

Figure 7-1. *The LM741, shown here, is still one of the most widely used op-amps.*

An ideal amplifier maintains a *linear relationship* between its input and its output, meaning that the output voltage values are a constant multiple of the input voltages over a wide range. This is

49

illustrated in Figure 7-2, where the lower curve is a duplicate of the upper curve, the only difference being that its amplitude is multiplied by a fixed amount (usually much greater than shown here). The ratio is properly known as the *gain* of the amplifier, usually represented with letter A (for amplification).

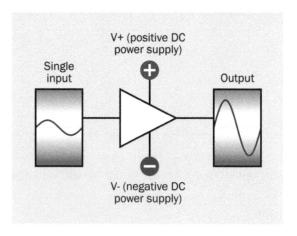

Figure 7-3. *The generic symbol for a single-input amplifier (not an op-amp), with the positive side of its power supply being equal and opposite in value to the negative side, and 0V being at the midpoint between them.*

- While the blue negative symbol is generally used throughout this encyclopedia to indicate 0V ground, it represents a voltage identified as V− in a dual voltage power supply, being equal in value but opposite in polarity to the positive side of the supply, V+. (Sometimes these voltages are indicated as V− and V+.)

The output from this imaginary generic amplifier is shown in the figure as a linear amplification of the input.

Dual Inputs

An op-amp has two inputs instead of one, and amplifies the voltage difference between them. Its symbol is shown in Figure 7-4. The upper input in this figure is held at 0V, midway between V+ and V−. Because the op-amp has so much gain, an accurate reproduction of its input would create an output exceeding the voltage of the power supply. Because this is not possible, the output tends to become saturated and consequently is *clipped* when it reaches its maximum value, as shown in the figure. The thumbnail graphs give only an approximate impression, as they are not drawn to the same scale.

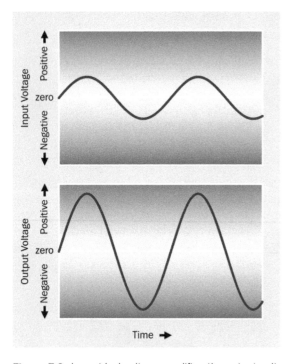

Figure 7-2. *In an ideal voltage amplifier, the output voltage will be a duplicate of the fluctuating input voltage, the only difference being that the amplitude of the output is multiplied by a fixed amount. This ratio is known as the* gain *of the amplifier.*

Figure 7-3 shows the triangular symbol for a generic single-input amplifier (not an op-amp). It may contain any number of components. The triangle almost always points from left to right, with its input on the left and its output on the right, and power attached above and below. This is often a *dual voltage* power supply, which is convenient for amplifying a signal that fluctuates above and below 0V. In some schematics, the power supply connections may not be shown, as they are assumed to exist.

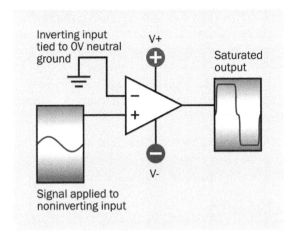

Figure 7-4. *An op-amp has so much gain, its output will tend to saturate, producing a square wave regardless of the shape of the input.*

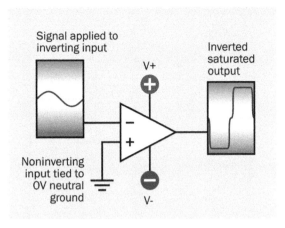

Figure 7-5. *When the incoming signal is applied to the inverting input of an op-amp while its noninverting input is held at 0V ground, the output is inverted.*

The small black plus and minus signs alongside the two inputs to the op-amp have nothing to do with the voltage supplied to the component. The "minus" input is properly referred to as the *inverting input* while the "plus" input is the *noninverting input*, in recognition of their functions.

The inputs are sometimes arranged with the minus above the plus, and sometimes with the plus above the minus. Schematics should be inspected carefully to note which arrangement is being used.

The positive and negative power connections to the op-amp may be omitted, but if shown, they always place V+ at the top, regardless of which way around the inputs are presented.

If a signal is applied to the noninverting input, while the inverting input is held at 0V ground, the op-amp provides an output in which the voltage is not inverted relative to the input.

If the input connections are swapped, so that the inverting input receives the incoming signal while the noninverting input is tied to 0V ground, the output from the op-amp is inverted (the gain remains the same). See Figure 7-5.

- An op-amp that is being used without any other components to moderate its output is functioning in *open loop* mode.

Negative Feedback

To create an output that is an accurate replica of the input, the op-amp must be brought under control with *negative feedback* to the input signal. This is illustrated in Figure 7-6. A resistor connects the output back to the inverting input, so that the input is automatically reduced to the point where the output is no longer saturated. The values of R1 and R2 will determine the gain of the op-amp, as explained in "How to Use It" on page 53. The op-amp is now functioning in its intended *closed loop* mode, meaning that the output is being tapped for feedback.

To obtain a linear output that is noninverted, connections are made as shown in Figure 7-7. The resistors form a voltage divider between the output and 0V ground, effectively increasing the comparison value on the inverting input.

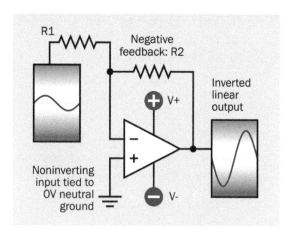

Figure 7-6. *A resistor applies negative feedback to the inverting input of an op-amp, and creates a linear output.*

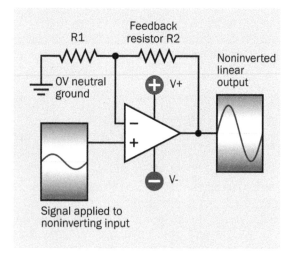

Figure 7-7. *Where the incoming signal is applied to the noninverting input, negative feedback is created by using a pair of resistors forming a voltage divider between the output and 0V ground.*

- Note that the gain of an op-amp is specific to a particular frequency range of AC signal. This is discussed in "How to Use It" on page 53.

Op-Amps and Comparators

A **comparator** can be regarded as a type of op-amp, and in fact an op-amp can be used as a comparator, comparing a variable DC voltage on one input with a reference voltage on another

input. However, the two types of components have diverged in design to the point where they should be considered separately. The distinction is sufficiently important to have prompted Texas Instruments to issue an Application Report in 2001 titled "Op Amp and Comparators—Don't Confuse Them!".

Differences in function are summarized in the previous entry discussing comparators (see "Differences from an Op-Amp" on page 42).

Variants

Because op-amps are mostly low-current devices, they are widely available in very small surface-mount formats, in addition to the through-hole DIP packages which used to be more common.

Many chips are available containing two or more op-amps. This is often expressed as the number of *channels* in the component. A dual chip contains two op-amps, while a quad chip contains four op-amps. Usually all the op-amps in a chip share the same power supply. Bipolar or CMOS transistors may be used.

Because op-amps are widely available in dual and quad packages, it's quite common for a circuit designer to have one op-amp in a chip "left over." The designer may be tempted to use that spare unit as a comparator instead of installing an additional chip. To address this situation, some manufacturers offer hybrid op-amp chips containing an additional comparator. The Texas Instruments TLV2303 and TLV2304 are examples.

Values

The op-amps derived from 1970s designs often tolerate a wide range of power-supply voltages. Plus-or-minus 5VDC to plus-or-minus 15VDC is a common range. Modern op-amps are available that run from as little as 1VDC to as much as 1,000VDC.

Op-amps are available for frequencies ranging from 5KHz all the way up to 1GHz.

A "classic" op-amp such as the LM741, which is still widely used, will operate with a power supply ranging from plus-or-minus 5VDC to plus-or-minus 22VDC. Its output is rated for up to 25mA, and its input impedance is at least 2MΩ. The most current it will draw from an input is around 0.5μA.

V_{IO} is the *input offset voltage*. In an ideal component, the output from an op-amp should be 0V when its inputs have a voltage difference of 0V. In practice, the output will be 0V when the inputs differ by the offset voltage. V_{IO} is likely to be no greater than a couple of mV, and negative feedback can compensate for the offset.

V_{ICR} is the *common mode voltage range*. This is the range of input voltages that the op-amp will tolerate. This can never be more than the positive power supply voltage and will often be less, depending on the types of transistors that are used on the input side. If an input voltage goes outside the common mode voltage range, the op-amp will stop functioning.

V_{IDR} is the *input differential voltage range*—the maximum permissible difference between peak positive and peak negative input voltages. This is often expressed as plus-or-minus the power supply voltage, or slightly less. Exceeding the range can have destructive consequences.

I_B is the *input bias current*, averaged over the two inputs. Most op-amps have extremely high input impedance and consequently use very low input currents.

Slew rate at unity gain is the rate of change of the output voltage caused by an instantaneous change on the input side, when the output of the op-amp is connected directly back to the inverting input (during operation in noninverting mode).

How to Use It

In addition to being an amplifier for AC signals, an op-amp can serve as an oscillator, filter, signal conditioner, actuator driver, current source, and voltage source. Many applications require some understanding of the complexities of mathematics describing alternating current, which are not included in this encyclopedia. Almost all the applications have a common starting point, however, which is to establish and control the gain of the feedback circuit.

Controlling the Gain

"A_{VOL} is the *open-loop voltage gain*, defined as the maximum voltage amplification that can be achieved when no feedback is applied from the output to an input. This remains constant until the AC frequency rises to a point known as the *breakover frequency*. If the frequency continues to rise, the maximum gain diminishes quite rapidly, until finally it terminates in 1:1 amplification at the *unity gain frequency*. This transition is shown by the orange line in Figure 7-8. The length of each purple line shows the frequencies which can be tolerated when the op-amp is used in closed-loop mode, and a negative feedback loop limits the gain. For example, where the gain is just 10:1, it can remain constant to just above 10KHz.

Note that both of the scales in this graph are logarithmic.

Calculating Amplification

So long as an op-amp is used within the boundaries of the graph, its voltage amplification can be controlled by choosing appropriate feedback and input resistors. If the op-amp is being used in noninverting mode, and R1 and R2 are placed as shown in Figure 7-7, the amplification ratio, A, is found approximately by the formula:

```
A = (approximately) 1 + (R2 / R1)
```

From this it can be seen that when R1 is very large compared with R2, the gain diminishes to near unity. If R1 becomes infinite and R2 is zero, the gain is exactly 1:1. This can be achieved by replacing R2 with a section of wire (theoretically of zero resistance) and omitting R1 entirely, as in Figure 7-9. In this configuration, the output from the op-amp should be identical with its input.

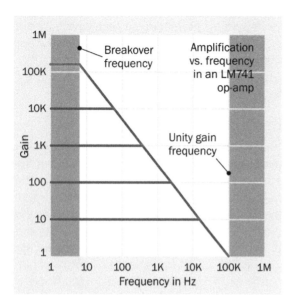

Figure 7-8. *Where each horizontal purple line meets the diagonal orange line, this is the maximum frequency that can be used without reduction in the maximum gain of an op-amp.*

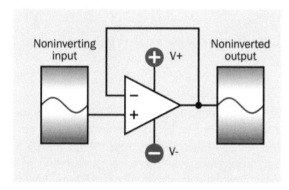

Figure 7-9. *While an op-amp is in noninverting mode, if the feedback resistor is replaced with a section of wire and the 0V ground connection is omitted entirely, the gain of the op-amp diminishes theoretically to 1:1.*

If the op-amp is being used in inverting mode, and R1 and R2 are placed as shown in Figure 7-6, then the voltage amplification ratio, A, is found approximately by the formula:

    A = (approximately) –(R2 / R1)

- Note the minus sign. In inverting mode, gain is expressed as a negative number.

- In a practical circuit, at the expected frequency, the amplification factor established by choice of resistors should be no more than 20.

- An inverting circuit has a relatively low input impedance. For this reason, in most applications, a noninverting circuit is preferred.

Unintentional DC Voltage Amplification

Although the op-amp is intended primarily as an AC signal voltage amplifier, it will also amplify a DC difference between the voltages on its inputs. In the upper section of Figure 7-10, a positive DC offset is inverted and amplified to the point where the output is forced to its negative limit, and the signal is lost, because its fluctuations have been overwhelmed by the positive offset. A coupling capacitor (shown in the lower section of the figure) removes the DC voltage while passing the AC signal. The appropriate capacitor value will depend on the frequency of the signal.

Low-Pass Filter

An op-amp can facilitate a very simple low-pass audio filter, just by adding a capacitor to the basic inverting circuit previously shown in Figure 7-6. The filter schematic is shown in Figure 7-11. Capacitor C1 is chosen with a value that passes higher audio frequencies and blocks lower audio frequencies. Because the gain of the basic inverting circuit is approximately –(R2 / R1), the op-amp functions normally when the impedance of C1 is blocking the low frequencies, forcing them to pass through R2. Higher frequencies, however, are able to bypass R2 through C1, lowering the effective resistance of the feedback section of the circuit, thus reducing its gain. This way, the power of the op-amp is greatly reduced for higher frequencies compared with lower frequencies. A passive RC circuit could achieve the same effect, but would attenuate the signal, while the op-amp circuit boosts part of it.

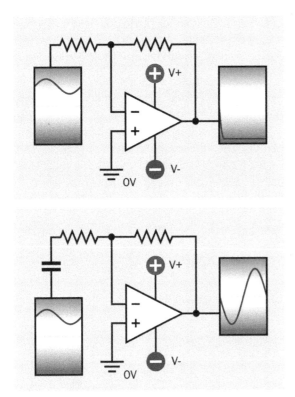

Figure 7-10. *The addition of a capacitor at the input of an op-amp is often necessary to prevent any DC voltage offset being amplified. In the upper section of this figure, a DC offset is large enough to force the inverted output to its negative limit, and the signal is completely lost.*

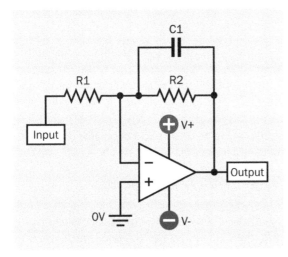

Figure 7-11. *A very basic low-pass filter, which works by allowing capacitor C1 to bypass resistor R2 at higher audio frequencies.*

High-Pass Filter

A simple high-pass filter can be created by adding a capacitor to the basic noninverting circuit previously shown in Figure 7-7. The filter schematic is shown in Figure 7-12. Once again capacitor C1 is chosen with a value that passes higher audio frequencies and blocks lower audio frequencies. Because the gain of the basic noninverting circuit is approximately $1 + (R2 / R1)$, the op-amp functions normally when the impedance of C1 is blocking the low frequencies, forcing them to pass through R1. Higher frequencies, however, are able to bypass R1 through C1, lowering the effective resistance of that section of the circuit, thus reducing the negative feedback and increasing the gain. This way, the power of the op-amp is increased for higher frequencies compared with lower frequencies. A passive RC circuit could achieve the same effect, but would attenuate the signal, while the op-amp circuit boosts part of it.

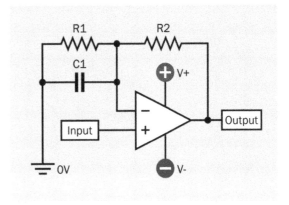

Figure 7-12. *A very basic high-pass filter, which works by allowing capacitor C1 to bypass resistor R1 at higher audio frequencies.*

Relaxation Oscillator

The schematic in Figure 7-13 is similar to the circuit shown in Figure 6-12 using a comparator. It functions as a *relaxation oscillator*, which is a form of *astable multivibrator*. The lower half of the circuit is a positive feedback loop that reinforces the output while the upper half of the circuit is charging the capacitor. Eventually the charge on the capacitor exceeds the voltage on the noninverting input of the op-amp, creating negative feedback that exceeds the positive feedback. The capacitor discharges and the cycle repeats. The component values in the figure should generate an output that runs at around 2Hz. Reducing the value of the capacitor will increase the frequency.

Single Power Source

A few op-amps are designed to work from single voltages, but they are a relatively small minority, and will clip the output signal if the input goes negative. Power supplies are readily available that provide multiple voltages such as +15VDC, 0V, and -15VDC. They are ideal for driving an op-amp—but may not be useful for any other components in the circuit. Can an op-amp that is designed for dual voltages be made to run from a single supply, such as 30VDC?

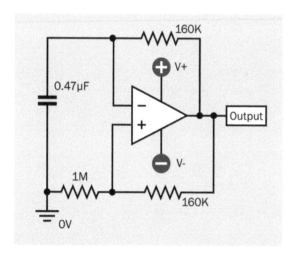

Figure 7-13. *A relaxation oscillator.*

This is relatively easy to do. The op-amp simply needs a potential difference to power its internal transistors, and 30VDC on the V+ pin with 0VDC on the V− pin will work just as well as +15VDC and -15VDC. However, referring back to Figure 7-6, if the op-amp is used in inverting mode, an intermediate voltage must be supplied to the noninverting input. Likewise, in noninverting mode, an intermediate voltage is necessary for one of the inputs, and must be half-way

between the extremes of the power supply. If the supply is +15VDC and -15VDC, the midpoint is 0V. If the supply is 30VDC and 0V, the midpoint is 15VDC.

Because the inputs of an op-amp have a very high impedance and draw negligible current, the intermediate voltage can be provided with a simple voltage divider, as shown in Figure 7-14, where R3 and R4 should be no greater than 100K each. Their exact values are not important, so long as they are equal.

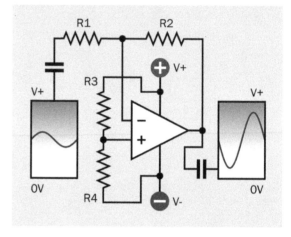

Figure 7-14. *A voltage divider, comprised of R3 and R4 in this schematic, can provide a voltage halfway between V+ and negative ground, enabling the op-amp to use just one power supply instead of two.*

A coupling capacitor should still be used on the input side, as shown, because there is no guarantee that the input signal will be centered precisely on 15V, and any offset will be amplified, potentially causing clipping of the signal. For similar reasons, a coupling capacitor is also added on the output side.

Offset Null Adjustment

Some op-amps provide two pins for *offset null adjustment*, which is a setup process to ensure that identical voltage on the two inputs will produce a null output. This is a way of compensating for any internal inconsistencies introduced during the manufacturing process.

To perform offset null adjustment, both input pins are connected directly to 0V ground, and the ends of a trimmer potentiometer (typically, 10K) are connected with the offset null pins, while the wiper of the potentiometer is centered and then connected with the negative power supply. The probes of a meter that is set to measure DC volts are placed between the output of the op-amp and 0V ground. The potentiometer is then adjusted until the meter shows a reading of 0VDC. A schematic is shown in Figure 7-15.

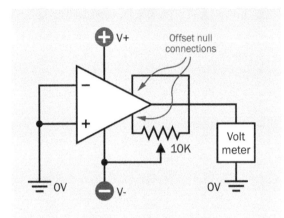

Figure 7-15. Connections for making an offset null adjustment to an op-amp that allows this procedure.

What Can Go Wrong

Power Supply Problems

Op-amps are especially vulnerable to reversed polarity in a power supply. If there is even a remote possibility of this occurring, a diode in series with one side of the supply can provide protection.

A more realistic concern is the destructive consequence of an input signal that exceeds the power supply voltage(s) of the op-amp. Even if the input is within the acceptable range, it can still cause permanent damage if it is applied before the op-amp powers up.

Bad Connection of Unused Sections

Multiple op-amps are often combined in a single package. If some of these "sections" remain unused, they will still receive power from the shared supply, and will attempt to function. If the inputs are left unconnected, they will pick up small stray voltages by capacitance or induction, and in the absence of negative feedback, the op-amp will create unpredictable outputs, consuming power and possibly interacting with other sections of the same chip. Figure 7-16 shows three incorrect options for addressing this problem, and one recommended option (derived from Texas Instruments Application Report SLOA067).

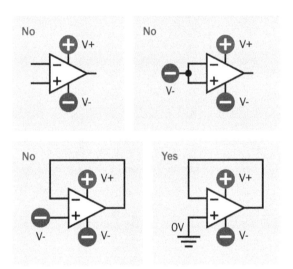

Figure 7-16. When multiple op-amps share a chip, one that is unused will still receive power from the shared supply. Its inputs must not be allowed to float, and must be connected to minimize activity and power consumption. Three common errors are shown here, with one recommended option. Note the distinction between 0V ground (0V) and negative power (V–).

Oscillating Output

The inputs of an op-amp are vulnerable to stray electromagnetic fields. If the conductors leading to and from the op-amp are relatively long, the output can couple capacitively with the input during voltage transitions, causing unwanted oscillations.

The commonly recommended solution to this problem is to add a 1µF bypass capacitor between the power supply and 0V ground. However, some manufacturers recommend alternatives such as introducing a very small amount of hysteresis, or reducing the value of input resistors to below 10K.

Confused Inputs

A schematic may show an op-amp with the non-inverting input above the inverting input, or vice versa. The only indication of this will be the plus and minus signs inside the chip, which can be extremely small and easily overlooked. For convenience in drawing a diagram, two op-amps in the same circuit may have their inputs shown in opposite configurations. Special care must be taken to verify that the inverting and noninverting input pins on a chip are correctly assigned.

digital potentiometer

A **digital potentiometer** is also known as a *digitally adjustable potentiometer*, a *digitally controlled potentiometer*, a *digitally programmed potentiometer* (with acronym *DPP*), a *digpot*, or a *digipot*. The terms are functionally interchangeable. Because the abbreviation *pot* is often used to describe an analog potentiometer, some people refer to digital potentiometers colloquially as *digital pots*. In printed documentation, the letters in *pot* may be capitalized. Because it is an abbreviation, not an acronym, it is not capitalized here.

Because this component enables digital control of a variable voltage, it is a *mixed signal device*. It is classified here as an analog chip because it primarily emulates the function of an analog device. It may be thought of as a form of *digital-analog converter*, although this encyclopedia does not have a section devoted to that type of component or to *analog-digital converters*, as their application is relatively specialized.

OTHER RELATED COMPONENTS

- **potentiometer** (see Volume 1)

What It Does

This component is an integrated circuit chip that emulates the function of an analog **potentiometer**. It is often described as being *programmable*, meaning that its internal resistance can be changed via a control input.

Digital potentiometers are particularly suited for use in conjunction with a *microcontroller*, which can control the internal resistance of the component. Possible applications include adjustment of the pulse width of an oscillator or multistable multivibrator (e.g., using the Control pin of a 555 **timer** chip); adjustment of the gain in an **op-amp**; specification of voltage delivered by a *voltage regulator*; and adjustment of a *bandpass filter*.

A digital potentiometer in combination with a microcontroller may also be used in conjunction with a pair of external buttons or a rotational en-

coder, to adjust the gain of an audio amplifier and for similar applications.

Advantages

A digital potentiometer offers significant advantages over an analog potentiometer:

- Reliability. The digital component may be rated for as many as a million cycles (each storing the wiper position in an internal memory location). An analog component may be capable of just a few thousand adjustment cycles.

- Digital interface.

- Elimination of long signal paths or cable runs. The digital potentiometer can be placed close to other chips, whereas an analog potentiometer often has to be some distance away to enable control by the end user. Reduction in the length of signal paths can reduce capacitive effects, while elimination of cable runs will reduce manufacturing costs.

- Reduction in size and weight compared with a manual potentiometer.

A digital potentiometer also has some disadvantages:

- Its internal resistance is somewhat affected by temperature.

- It is not usually capable of passing significant current. Few chips can sink or source more than 20mA at the output, and 1mA is common. The output is primarily intended for connection with other solid-state components that have high impedance.

- Users may prefer the immediacy and tactile feel of a knob attached to an analog potentiometer, rather than a pair of buttons or a rotational encoder.

How It Works

A digital potentiometer changes the point at which a connection is made along a *ladder* of many fixed resistors connected in series inside the chip. Each end of the ladder, and each intersection between two adjacent resistors, is known as a *tap*. The pin that can connect with any of the taps is referred to as the *wiper*, because it emulates the function of a wiper in an analog potentiometer. In reality, a digital potentiometer does not contain a wiper or any other moving parts.

A fully featured digital potentiometer allows access to each end of the ladder through two pins that are often labeled "high" and "low," even though they are functionally interchangeable (except in the case of a component that simulates a logarithmic taper, as described later). The "low" end of the ladder is sometimes numbered 0. In this case, if there are n resistors, the "high" end of the ladder will be numbered n. Alternatively, if the "low" end of the ladder is numbered 1, and there are n resistors, the "high" end will be

numbered $n+1$. This principle is illustrated in Figure 8-1.

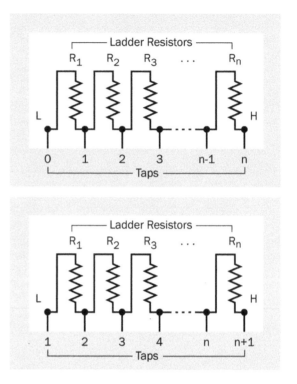

Figure 8-1. *Available wiper connections to a resistor ladder inside a digital potentiometer, showing two numbering systems that may be used.*

The "low" pin on a digital potentiometer may be identified as L, or A, or R_L, or P_A in a datasheet, while the "high" pin may be identified as H, or B, or R_H, or P_B and the pin that accesses the wiper is typically identified as W, or R_W, or P_W. Letters L, H, and W are used below. Although the L and H pins are functionally interchangeable, their labels are useful to identify which direction the W connection will move in response to an external signal.

Digital potentiometers are available with as few as 4 or as many as 1,024 taps, but common values are 32, 64, 128, or 256 taps, with 256 being the most common.

No specific schematic symbol represents a digital potentiometer. Often the component is shown as an analog potentiometer symbol inside a box that has a part number, as suggested in

Figure 8-2. Control pins and the power supply may be omitted if the schematic is just intended to show logical connections. Alternatively, if the digital potentiometer is depicted in a schematic where it is connected with other components such as a microcontroller, multiple pins and functions may be included, as shown in Figure 8-3. The pins additional to L, H, and W are explained below.

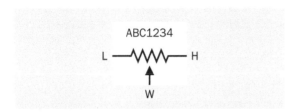

Figure 8-2. *There is no single specific symbol to represent a digital potentiometer. It may be shown using an analog potentiometer symbol in a box with a part number, as suggested here, where power connections and additional pins are omitted for clarity.*

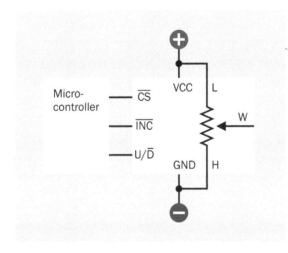

Figure 8-3. *If a digital potentiometer appears in a schematic where it is connected with other components such as a microcontroller, additional pins and functions may be indicated. This generic representation of a digital potentiometer shows some of the functions that can be included.*

Variants

A *dual* digital potentiometer contains two complete units, while a *quad* contains four. Triples

exist but are relatively uncommon. A few chips contain six potentiometers. Multiple digital potentiometers in a chip can be used as the digital equivalent of *ganged* analog potentiometers, for simultaneous synchronized adjustment of multiple inputs in an audio system (two channels in a stereo amplifier, or more in a surround-sound system).

The pinouts of a sophisticated quad digital potentiometer chip are shown in Figure 8-4. Other quad chips have different pinouts and capabilities; there is no standardized format as there is with digital logic chips. In this example, the high/low states of Address 0 and Address 1 select one of the four internal resistor ladders, numbered 0 through 3. The Chip Select pin makes the whole chip either active or inactive. The Write Protect pin disables writing to the internal wiper memory. The Serial Clock pin inputs a reference pulse stream to which the serial input data must be synchronized. The Hold pin pauses the chip while data is being transmitted, allowing the data transmission to be resumed subsequently. The NC pins have no connection.

Volatile and Nonvolatile Memory

Any type of digital potentiometer requires memory to store its current wiper position, and this memory may be volatile or nonvolatile. Nonvolatile memory may be indicated in a datasheet by the term *NV*.

A digital potentiometer with volatile memory will typically reset its wiper to a center-tap position if power is disconnected and then restored. A digital potentiometer with nonvolatile memory will usually restore the most recently used wiper position, provided the chip is fully powered down and then fully powered up without glitches in the supply. If a microcontroller is being used to control the digital potentiometer, it can store the most recent resistance value in its own nonvolatile memory, in which case the type of memory in the potentiometer becomes irrelevant.

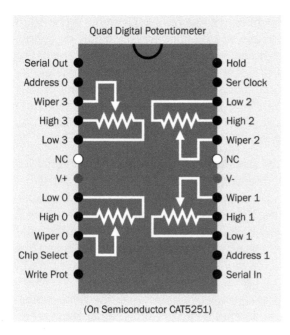

(On Semiconductor CAT5251)

Figure 8-4. *Pinouts of a sophisticated quad digital potentiometer chip. Other chips will have different pinouts and capabilities. This example is available in surface-mount formats only. See text for details.*

Taper

Digital potentiometers are available with *linear taper* or *logarithmic taper*. In the former, each resistor in the ladder has the same value. In the latter, values are chosen so that the cumulative resistance between the wiper and the L end of the ladder increases geometrically as the wiper steps toward the H end of the ladder. This is useful in audio applications where sound intensity that increases exponentially may seem to increase linearly when perceived by the human ear.

A microcontroller can emulate logarithmic steps by skipping some taps in the ladder in a digital potentiometer, but this will result in fewer increments and lower precision.

Data Transfer

Digital potentiometers are mostly designed to use one of three serial protocols:

- *SPI*. This acronym is derived from *serial peripheral interface*, a term trademarked by Motorola but now used generically. The standard is adapted in various radically different ways among digital potentiometers.

- *I2C*. More correctly printed as *I²C* and properly pronounced "I squared C," this acronym is derived from the term *inter-integrated circuit*. Developed by Philips in the 1990s, it is a relatively slow-speed bus-communication protocol (up to 400kbps or 1Mbps in its basic form). It is built into some microcontrollers. The standard is more uniformly and rigorously defined than SPI.

- *Up/down*, also sometimes known as *push-button* or *increment/decrement* protocol.

Both SPI and I2C are supported by many microcontrollers, including the Atmel AVR at the heart of the Arduino.

These three systems for controlling a digital potentiometer are described in more detail in the following sections.

SPI

This is the most widely used protocol, but when reading datasheets, a lot of care must be taken to determine how it varies in each case.

The Microchip 4131-503, shown in Figure 8-5, uses SPI protocol. It contains 128 resistors and can be powered by 1.8VDC to 5.5VDC.

The one feature that all versions of SPI have in common is that a series of high/low pulses is interpreted by the chip as a set of bits whose value defines a tap point in the resistor ladder. In computer terminology, every tap point has an *address*. The incoming bits define the address, after which the status of an additional input pin can tell the chip to move the wiper to that location.

Typically, there will be a chip select pin, identified as CS; a serial data input pin, identified as SDA, SI, DIN, or a similar acronym; and a serial clock pin, identified as SCL, SCLK, or SCK, which must receive a stream of pulses to which the high/low data input pulses must be synchronized. In addition, the SPI protocol allows bidirectional (duplex) serial communication. Only a minority of digital potentiometers make use of this capability, but where it exists, a serial data output pin may be labeled SDO. Alternatively, one pin may be multiplexed to enable both input and output, in which case it may be labeled SDI/SDO.

Figure 8-5. *This digital potentiometer uses SPI protocol. See text for details.*

If a pin is active-low, a bar (a horizontal line) will be printed above its acronym.

The most common type of digital potentiometer has 255 resistors and therefore 256 tap points, which have addresses numbered 0 through 255, each of which can be specified by a sequence of eight data bits constituting one byte. However, a different coding system will be applied in chips that have a different number of taps. In a 32-tap component, for instance, data is still sent in groups of eight bits, but only the first five bits define a *tap address*, while the remaining three are interpreted as commands to the chip.

Most 256-tap chips use an SPI protocol in which two eight-bit bytes are sent, the first being interpreted by the chip as a command, while the second specifies a tap address. Each manufacturer may use a different set of command codes, and these will vary among chips even from the same manufacturer.

Most commonly, three wires are used for data transmission and control (causing these chips to be described as *3-wire* programmable potentiometers).

CS is usually, but not always, pulled low to activate the digital potentiometer for input. A series of low or high states is then applied to the data-input pin. Each time the clock input changes state (usually on the rising edge of the clock pulse) the state of the data input is copied to a **shift register** inside the chip. After all the bits have been clocked in, CS can change from low to high, causing the contents of the shift register to be copied into a decoder section of the chip. The first bit received becomes the most significant bit in the decoder. The value of the eight bits is decoded, and the chip connects the W pin directly to the corresponding tap along the ladder of 255 internal resistors.

I2C Protocol

The I2C specification is controlled by NXP Semiconductors (formerly Philips), but can be used in commercial products without paying licensing fees. Only two transmission lines are required: one carrying a clock signal, the other allowing bidirectional data transfer synchronized with the clock (although many digital potentiometers use the I2C connection only to receive data). The pins are likely to be identified by the same acronyms as the pins on a chip that uses SPI protocol.

As in SPI, a command byte is followed by a data byte, although the command set differs from that of SPI and will also differ among various I2C chips. Full implementation of I2C allows multiple devices to share a single bus, but this capability may remain unused.

Up/Down Protocol

This simpler, asynchronous protocol does not require a clock input. The chip will respond to data pulses that are received at any speed (up to its maximum speed), and the pulse widths can be inconsistent.

Each pulse moves the wiper connection one step up or down the ladder. While this has the advantage of simplicity, the taps are not addressable, and consequently the wiper cannot skip to any tap without passing through intervening taps incrementally. This is not an inconvenience when the potentiometer controls audio gain, which is a primary application.

In some chips, an increment pin, usually labeled INC, receives pulses while the high or low state of a second pin, usually labeled U/D, determines whether each pulse will step the wiper up the ladder or down the ladder.

In other chips, pulses to an Up pin will step the wiper up the ladder, while pulses to a Down pin will step the wiper down the ladder.

Either of these chip designs can be referred to as a *two-wire* type. If an additional chip-select pin is included (labeled CS on datasheets), this type of digital potentiometer can be referred to as a *three-wire* type. The chip select pin is likely to be active-low, meaning that so long as it has a high state, the chip will ignore incoming signals.

The CAT5114 shown in Figure 8-6 uses an U/D pin. It contains 31 resistors, is available in 8-pin DIP or surface-mount formats, and can be powered by 2.5VDC to 6VDC. Each of its logic inputs draws only 10µA.

In six-pin chips the INC pin is omitted, one of the H, L, or W pins will be omitted, and the U/D pin will function differently. When CS is pulled low, the chip checks the state of the U/D pin. If it is high, the chip goes into increment mode; if it is low, the chip goes into decrement mode. So long as CS remains low, each transition of the U/D pin from low to high will either increment or decrement the wiper position, depending on the mode that was sensed initially. When CS goes high, further transitions on the U/D pin will be ignored until CS goes low again, at which point the procedure repeats.

Figure 8-6. *This digital potentiometer contains 31 resistors and uses the simplest up/down protocol to step from one tap to the next.*

The chip does not provide any feedback regarding the position of its wiper, and consequently a control device such as a microcontroller cannot know the current wiper position. If the chip has nonvolatile memory (as is the case in many up/down digital potentiometers), it will resume its previous wiper location at power-up, but here again a control device will have difficulty determining what that position is. Therefore, in its basic form, an up/down chip is only appropriate for simple tasks, especially in response to up/down pushbuttons.

Other Control Systems

A few digital potentiometers use a parallel interface. Because they are relatively uncommon, they are not included here.

Connections and Modes

Some variants of digital potentiometers minimize the chip size and number of connections by limiting accessibility to the internal resistor ladder. In a chip designed to function in *rheostat mode*, the W pin is eliminated and the chip moves an internal connection point to change the resistance between the H and L pins.

In some variants, the low end of the ladder is permanently, internally connected with ground, and the L pin is omitted. In other variants, one end of the ladder is unconnected inside the chip.

A chip designed to function in *voltage divider mode* will include all three pins—H, L, and W—except in some instances where the low end of the ladder is grounded internally.

Variants are shown in Figures 8-7, 8-8, 8-9, and 8-10. Because some pins may be omitted, and there is no standardization of function among the pins that do exist, circuits and chips must be examined carefully prior to use.

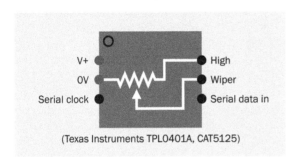

(Texas Instruments TPL0401A, CAT5125)

Figure 8-7. *Some digital potentiometers minimize chip size and provide specialized functionality by eliminating pins. In the variant shown here, the W pin provides a voltage between H and an internal ground connection. The chip is controlled via I2C serial protocol.*

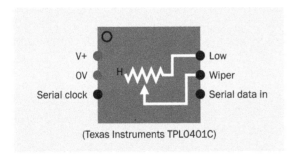

(Texas Instruments TPL0401C)

Figure 8-8. *In this variant, the H end of the internal resistor ladder is allowed to float inside the chip, and the digital potentiometer functions as a rheostat. The chip is controlled via I2C serial protocol.*

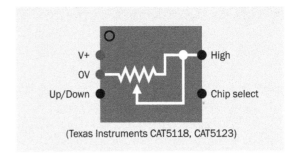

(Texas Instruments CAT5118, CAT5123)

Figure 8-9. *This variant provides a variable resistance between the H pin and an internal connection with negative ground. Pin 5 is omitted. The chip is controlled by up/down pulses.*

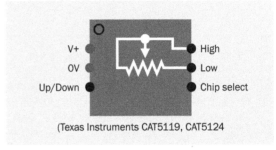

(Texas Instruments CAT5119, CAT5124

Figure 8-10. *This variant provides a variable resistance between H and L pins, without allowing either end of the resistor ladder to float. The W pin is omitted, as the wiper is tied internally to the H pin. The chips listed are controlled by up/down pulses.*

Values

A primary limitation of digital potentiometers is that they cannot withstand significant current. This may prevent them from being substituted for an analog potentiometer unless changes are made in the circuit. H, L, and W pins are usually unable to source or sink continuous, sustained current exceeding 20mA.

Wiper resistance is the resistance that is added internally by the wiper. This is nontrivial; it is often around 100Ω, and can be as high as 200Ω.

Typical end-to-end resistance of the ladder of internal resistors may range from 1K to 100K. Values of 1K, 10K, and 100K are common.

While the number of taps is likely to be a power of two in chips where the taps are addressable, up/down chips are not so constrained and may contain, for example, 100 taps.

The end-to-end resistance of a whole ladder may vary by as much as 20% from one sample of a chip to the next. Among resistor ladders in digital potentiometers sharing the same chip (i.e., in dual or quad chips) the variation will be much smaller.

Almost all digital potentiometers are designed for a supply voltage of 5V or less. The H and L pins are not sensitive to polarity, but the voltage applied to either of them must not exceed the supply voltage.

How to Use It

While most microcontrollers contain one or more analog-digital converters that change an analog *input* to an internal numeric value, a microcontroller cannot create an analog *output*. A digital potentiometer adds this functionality, although applications will be restricted by its limitation on current.

An up/down digital potentiometer can be controlled directly by a pair of **pushbuttons**, one of which will increase the resistance value while the other will reduce it. The pushbuttons must be *debounced* when used in this way. An alternative to pushbuttons is a **rotational encoder**, which emits a stream of pulses when its shaft is turned. In this case, an intermediate component (probably a microcontroller) will usually interpret the pulse stream and change it to a format that the digital potentiometer can understand.

Where a digital potentiometer is used in audio applications, it should be of the type that moves the wiper connection from one tap to another during a *zero crossing* of the audio signal (i.e., that is, at the moment when the AC input signal passes through 0V on its way from positive to negative or negative to positive). This suppresses the "click" that otherwise tends to occur during

switching. Potentiometers with this feature may include phrases such as "glitch free" in their datasheets.

Digital potentiometers that are intended primarily for audio applications often have 32 taps spaced at intervals of 2dB. This will be sufficient to satisfy most listeners.

Achieving Higher Resolution

For sensitive applications where a resolution with more than 1,024 steps is required, multiple digital potentiometers with different step values can be combined. One way of doing this is shown in Figure 8-11. In this circuit, the wipers of P2 and P3 must be moved in identical steps, so that the total resistance between the positive power supply and negative ground remains constant. These two potentiometers could be contained in a dual chip, and would receive identical up/down commands. P1 is at the center of the voltage divider formed by P2 and P3, and is adjusted separately to "fine tune" the output voltage that is sensed at point A.

If all three of the potentiometers in this circuit contain 100 taps, a combined total of 10,000 resistance steps will be possible.

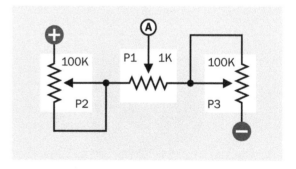

Figure 8-11. *If all three digital potentiometers in this schematic have 100 taps, and the wipers of P2 and P3 are moved in synchronization, the voltage measured at point A can have a high resolution of up to 10,000 steps.*

What Can Go Wrong

Noise and Bad Inputs

Because a digital potentiometer is capable of receiving data at speeds as high as 1MHz, it is sensitive to brief input or power fluctuations, and can misinterpret them as instructions to move the wiper—or can misinterpret them as command codes, in a component using SPI or I2C serial protocol.

To minimize noise in the power supply, some manufacturers recommend installing a 0.1μF capacitor as close as possible to the power supply pin of the component. In addition, it is obviously important to provide clean input signals. This means thorough debouncing of any electromechanical switch or pushbutton inputs.

Wrong Chip

The wide diversity of input protocols and pinouts creates many opportunities for installation error.

Up/down, SPI, and I2C protocol require totally different pulse streams. Many manufacturers offer components that are distinguished from each other by just one or two digits in their part numbers, yet have radically different functionality.

If more than one specific type of digital potentiometer may be used during circuit development, they should be stored carefully to avoid inadvertant substitutions. Using the wrong chip may be particularly confusing in that an inappropriate input protocol will still produce some results, although not those which were intended.

Controller and Chip Out of Sync

As noted in the discussion of data transmission protocols, most digital potentiometers are not capable of providing feedback to confirm the position of the internal wiper. A designer may wish to include a power-up routine which establishes the state of the digital potentiometer by resetting it to a known position, at one end of its scale or the other.

Nonlinear Effects

While the end-to-end resistance of the resistor ladder inside a digital potentiometer is not likely to be affected significantly by changes in temperature, the resistance at the wiper is more heat sensitive.

In an up/down chip, there can be differential errors between incremental and decremental modes. In other words, if a tap is reached by stepping up to it incrementally, the resistance between the W pin and H or L may not be quite the same as if the same tap is reached by stepping down to it decrementally. The difference may not be significant, but may be puzzling for those who are unfamiliar with this phenomenon.

Some differences may be found among resistors in a ladder. That is, in a supposedly linear digital potentiometer, each resistor may differ in value slightly from the next.

Data Transfer Too Fast

When using a microcontroller to send data to a digital potentiometer, a small delay may be necessary between pulses, depending on the microcontroller's clock speed. A digital potentiometer may require a minimum pulse duration of 500ns. Check the manufacturer's datasheet for details.

timer

S

A device that creates a single timed pulse, or a series of timed pulses with timed intervals between them, is properly known as a *multivibrator*, although the generic term **timer** has become much more common and is used here.

Three types of multivibrator exist: *astable*, *monostable*, and *bistable*. The behavior of *astable multivibrators* and *monostable multivibrators* is described in detail in this entry. A timer chip can also be made to function as a bistable multivibrator. This is described briefly below, but it is not a designed function of a timer. The primary discussion of bistable multivibrators will be found in the entry of this encyclopedia dealing with **flip-flops**.

OTHER RELATED COMPONENTS

- **flip-flop** (See Chapter 11)

What It Does

A *monostable* timer emits a single timed pulse of fixed length in response to a triggering input that is usually of shorter duration. Many monostable timers are also capable of running in *astable* mode, in which the timer spontaneously emits an ongoing stream of timed pulses with timed gaps between them. A dual-mode timer can run in either mode, determined either by external components attached to it, or (less commonly) by changing the status of a mode selection pin.

Monostable Mode

In monostable mode, the timer emits a pulse in response to a change from high to low voltage (or, less commonly, from low to high voltage) at a *trigger pin*. Most timers respond to a *voltage level* at the trigger pin, but some are insensitive to any persistent pin state and only respond to a *voltage transition*. This is known as *edge triggering*.

The pulse generated by the timer may consist of a change from low to high (or, less commonly, from high to low) at an output pin. The length of the pulse will be determined by external components, and is independent of the duration of the triggering event, although in some cases, an output pulse may be prolonged by *retriggering* the timer prematurely. This is discussed below.

At the end of the output pulse, the timer reverts to its quiescent state, and remains inactive until it is triggered again.

A monostable timer can control the duration of an event, such as the time for which a light remains on after it has been triggered by a motion sensor. Alternatively, the timer can impose a delay, such as the interval during which a paper towel dispenser refuses to respond after a towel has been dispensed. A timer can also be useful to generate a clean pulse in response to an unstable or noisy input, as from a manually operated pushbutton.

Astable Mode

In astable mode, a timer will generally trigger itself as soon as power is connected, without any need for an external stimulus. However, the output can be suppressed by applying an appropriate voltage to a *reset pin*.

External components will determine the duration of each pulse and the gap between it and the next pulse. The pulse stream can be slow enough to control the flashing of a turn signal in a 1980s automobile, or fast enough to control the bit rate in a data stream from a computer.

Modern timer circuits are often incorporated in chips that have other purposes. The flashing of a turn signal in a modern car, for instance, is now likely to be timed by a microcontroller that manages many other functions. Still, chips that are purely designed as timers remain widely used and are very commonly available in numerous through-hole and surface-mount formats.

How It Works

The duration of a single pulse in monostable mode, or the frequency of pulses in astable mode, is most commonly determined by an external *RC network* consisting of a resistor in series with a capacitor. The charging time of the capacitor is determined by its own size and by the value of the resistor. The discharge time will be determined in the same way. A **comparator** inside the timer is often used to detect when the potential on the capacitor reaches a *reference voltage* that is established by a voltage divider inside the chip.

Variants

The 555 Timer

An eight-pin integrated circuit manufactured by Signetics under part number 555 was the world's first fully functioned timer chip, introduced in 1972. It combined two comparators with a **flip-flop** (see Chapter 11) to enable great versatility while maintaining excellent stability over a wide range of supply voltages and operating temper-

atures. Subsequent timers have been heavily influenced by this design. A typical 555 timer chip is shown in Figure 9-1.

Figure 9-1. *A typical 555 timer chip. Functionally identical versions in which the "555" identifier is preceded or followed by different letter combinations are available from many different manufacturers.*

The 555 was designed by one individual, Hans Camenzind, working as an independent consultant for Signetics. According to a transcript of an interview with Camenzind maintained online at the Transistor Museum, "There was nothing like it at the time. You had to use quite a few discrete components—a comparator, a Zener diode or even two. It was not a simple circuit."

The 555 timer quickly became the most widely used chip in the world, and was still selling an annual estimated 1 billion units three decades after its introduction. It has been used in spacecraft, in intermittent windshield wiper controllers, in the early Apple II (to flash the cursor), and in children's toys. Like many chips of its era, its design was unprotected by patents, allowing it to be copied by numerous manufacturers.

The initial version was built around **bipolar transistors**, and consequently is referred to as the *bipolar version* or (more often) the *TTL version*, this being a reference to *transistor-transistor logic* protocol. Within a few years, CMOS versions based around **MOSFETs** were developed. They reduced the ability of the chip to sink or source current at its output pin, but consumed far less power, making them better suited to battery-operated products. The CMOS versions were and still are pin-compatible with the original bipolar version, both in through-hole and surface-mount formats. Their timing parameters are usually the same.

555 Monostable Operation

The internal functionality of a 555 timer wired to run in monostable mode is illustrated in Figure 9-2 with the chip seen from above. The pins are identified in datasheets by the names shown. To assist in visualizing the behavior of the chip, this figure represents the internal flip-flop as a switch which can be moved by either of two internal comparators, or by an input from the Reset pin.

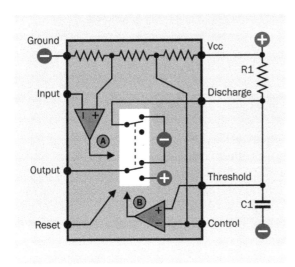

Figure 9-2. *The internal functions of a 555 chip, with its flip-flop represented as a switch that can be moved by either of two comparators, or by a low voltage on the Reset pin. An external resistor and capacitor, shown as R1 and C1, cause the timer to run in monostable (one-shot) mode, generating a single high pulse when the state of the Input pin is pulled from high to low.*

Inside the chip, three resistances of 5K each are connected between V+ (positive supply voltage) and negative ground. It has been suggested that the part number of the 555 chip was derived from these three 5K resistors, but Hans Camenzind has

pointed out that Signetics was already using three-digit part numbers beginning with the number 5, and probably chose the 555 part number because the sales department had high expectations for the chip and wanted its number to be easily memorable. (A similar rationale explains the part number of the 2N2222 transistor.)

The resistances inside the timer function as a *voltage divider*, providing a reference of 1/3 of V+ to the noninverting pin of Comparator A and 2/3 of V+ to the inverting pin of Comparator B. (See Chapter 6 for an explanation of the functioning of comparators.)

When power is initially supplied to the timer, if the Input pin is at a high state, Comparator A has a low output, and the flip-flop remains in its "up" position, allowing the Output pin to remain in a low state. The flip-flop also grounds the lower end of R1, which prevents any charge from accumulating on capacitor C1.

If the state of the Input pin is pulled down externally to a voltage less than 1/3 of V+, Comparator A now creates a high output that changes the flip-flop to its "down" position, sending a high signal out through the Output pin. At the same time, C1 is no longer grounded, and begins to charge at a rate determined by its own size and by the value of R1. When the charge on the capacitor exceeds 2/3 of V+, it activates Comparator B, which forces the flip-flop into its "up" position. The Output pin goes low, C1 discharges itself into the Discharge pin, and the timer's cycle is at an end.

The low voltage on the Input pin of the timer must end before the end of the output cycle. If the voltage on the Input pin remains low, it will re-trigger the timer, prolonging the output pulse.

A pullup resistor may be used on the Input pin to avoid false triggering, especially if an external electromechanical **switch** or **pushbutton** is used to pull down the Input pin voltage.

The Reset pin should normally be held high, either by being connected directly to positive supply voltage (if the reset function will not be needed) or by using a pullup resistor. If the Reset pin is pulled low, this will always interrupt an output pulse regardless of the timer's current status.

If a voltage higher or lower than 2/3 of V+ is applied to the Control pin, this will change the reference voltage on Comparator B, which determines when the charging cycle of C1 ends and the discharge cycle begins. A lower reference voltage will shorten each output pulse by allowing a lower charge limit for C1. If the control voltage drops to 1/3 of V+ (or less), the capacitor will not charge at all, and the pulse length will diminish to zero. If the control voltage rises to become equal to V+, the capacitor will never quite reach that level, and the pulse length will become infinite. A workable range for the control voltage is therefore 40% to 90% of V+.

Because the Control pin is an input to the chip, it should be grounded through a 0.01µF ceramic capacitor if it will not be used. This is especially important in CMOS versions of the timer.

A defect of the bipolar 555 is that it creates a *voltage spike* when its Output pin changes state. If it will be sharing a circuit with sensitive components, a 0.01µF bypass capacitor should be added as closely as possible between the V+ pin and negative ground. The voltage-spike problem was largely resolved by the CMOS 555.

555 Astable Operation

In Figure 9-3, the 555 timer chip is shown with external components and connections to run it in astable mode. The pin names remain the same but have been omitted from this diagram because of limited space. The labeling of the two external resistors and capacitor as R1, R2, and C1 is universal in datasheets and manufacturers' documentation.

When the timer is powered up initially, capacitor C1 has not yet accumulated any charge. Consequently, the state of the Threshold pin is low. But the Threshold pin is connected externally with the Input pin, for astable operation. Consequently, the Input pin is low, which forces the flip-flop into its "down" state, creating a high output. This happens almost instantaneously.

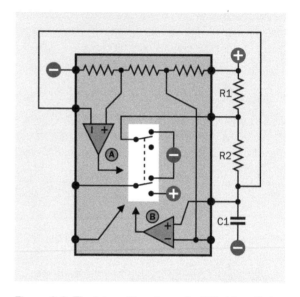

Figure 9-3. *The internal functions of a 555 chip, with two external resistors and a capacitor wired to run the timer in astable (free-running) mode.*

While the flip-flop is "down," the Discharge pin is not grounded, and current flowing through R1 and R2 begins to charge the capacitor. When the charge exceeds 2/3 of positive supply voltage, Comparator B forces the flip-flop into its "up" position. This ends the high pulse on the Output pin, and starts to drain the charge from the capacitor through R2, into the Discharge pin. However, the voltage on the capacitor is still being shared by the Input pin, and when it diminishes to 1/3 of V+, the Input pin reactivates Comparator A, starting the cycle over again.

The functions of the Reset and Control pins are the same as in monostable mode. Because voltage applied to the Control pin changes the length of each pulse and the gaps between pulses, it has the effect of adjusting the frequency of the output in astable mode.

When power is first connected to the timer, C1 must initially charge from an assumed state of zero potential to 2/3 V+. Because subsequent cycles will begin when the capacitor is at 1/3 V+, the first high output pulse from the timer will be slightly longer than subsequent output pulses. This is unimportant in most applications, especially because the rate at which a capacitor accumulates charge is greater when beginning from 0V than when it has reached 1/3 V+. Still, the longer initial pulse can be noticeable when the timer is running slowly.

Because the capacitor charges through R1 and R2 in series, but discharges only through R2, the length of each positive output pulse in astable mode is always greater than the gap between pulses. Two strategies have been used to overcome this limitation. See "Separate Control of High and Low Output Times" on page 80.

556 Timer

The 556 consists of two 555 bipolar-type timers in one package. An example of the chip is shown in Figure 9-4. The pinouts are shown in Figure 9-5. Although 556 timers have become relatively uncommon compared with the 555, they are still being manufactured in through-hole and surface-mount versions by companies such as Texas Instruments and STMicroelectronics, under part numbers such as NA556, NE556, SA556, and SE556 (with various letters or letter pairs appended). Each timer in the chip has its own set of inputs and outputs, but the timers share the same V+ and ground voltages.

558 Timer

This 16-pin chip is now uncommon, and many versions have become obsolete. It has been identified by a part number such as NE558 although different prefix letters may be used. The NTE926, shown in Figure 9-6, is actually a 558 timer.

The chip contains four 555 timers sharing a common power supply, common ground, and common control-pin input. For each internal timer, the Threshold and Discharge functions are con-

nected internally, so that the timers can only be used in one-shot mode. However, one timer can trigger another at the end of its cycle, and the second timer can then retrigger the first, to create an astable effect.

Each timer is edge-triggered by a voltage transition (from high to low), instead of being sensitive to a voltage level, as is the case with a 555 timer. Consequently the timers in the 558 chip are insensitive to a constant (DC) voltage.

Figure 9-4. *An example of the 556 timer chip.*

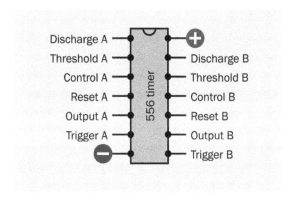

Figure 9-5. *The 556 timer contains two separate 555 timers sharing the same power supply and ground. The pin functions for timer A and timer B are shown here.*

The output from each timer is an open collector, and therefore requires an external pullup resistor. Each output is capable of sinking up to 100mA.

Figure 9-6. *The NTE926 is a 558 timer chip.*

CMOS 555 Timer

While the part numbers of many CMOS versions are significantly different from part numbers of the bipolar versions, in some instances the CMOS numbers are only distinguished by a couple of initial letters. The ST Microelectronics TS555 series and Texas Instruments TLC555 series, for example, use MOSFETs internally. The ST Microelectronics SE555 series and Texas Instruments SA555 series use bipolar transistors internally.

One way to distinguish between the two types, when searching a website maintained by a parts supplier, is to begin by looking generically for a "555 timer" and then add a search filter to show chips either with a minimum power supply of 3VDC (which will be CMOS) or with a minimum power suply of 4.5VDC (which will be bipolar).

CMOS versions of the 555 timer do not create the power spike that is characteristic of the bipolar versions during output transitions. The CMOS chips can also be powered by a lower voltage (3VDC, or 2VDC in some cases), and will draw significantly less current in their quiescent state. They also require very little current for threshold, trigger, and reset functions.

The wiring of external resistors and capacitors to the CMOS version of the chip, and the internal voltage levels as a fraction of V+, are identical to

the original 555 timer. Pin functions are likewise identical. The only disadvantages of CMOS versions are their greater vulnerability to static discharge, and their lower output currents. The TLC555, for instance, will source only 15mA (although it can sink 10 times that amount). Other manufacturers have different specifications, and datasheets should be checked carefully.

5555 Timer

The 5555 contains a digital **counter** that enables it to time very long periods. Its full part number is 74HC5555 or 74HCT5555, although these numbers may be preceded or followed by letter combinations identifying the manufacturer. It is not pin-compatible with a 555 timer.

Two input pins are provided, one to trigger the timer on a rising edge, the other on a falling edge, of the input pulse. The inputs are Schmitt-triggered.

The timer is rated for 1Hz to 1MHz (using an external resistor and external capacitor). The counter section can divide the pulse frequency by values ranging from 2 to 256. For longer timed periods, different settings on the digital control pins will divide the frequency by values ranging from 2^{17} through 2^{24} (131,072 through 16,777,216). This enables the timer to achieve a theoretical pulse length lasting for more than 190 days. The timer will accept a clock input from an external oscillator to achieve better accuracy than is available with a resistor-capacitor timing circuit.

7555 Timer

This 8-pin chip is a CMOS version of the 555 timer, manufactured by companies such as Maxim Integrated Products and Advanced Linear Devices. Its characteristics are similar to those of CMOS 555 timers listed above, and the pinouts are the same.

7556 Timer

This 14-pin chip contains two 7555 timers, sharing common power supply and ground connections. Pinouts are the same as for the original 556 timer, as shown in Figure 9-5.

4047B Timer

This 14-pin CMOS chip was introduced in an effort to address some of the quirks of the 555 timer while also providing additional features. It runs in either monostable or astable mode, selectable by holding one input pin high or another input pin low. In astable mode, its duty cycle is fixed at approximately 50%, a single resistor being used for both charging and discharging the timing capacitor. An additional "oscillator" output runs twice as fast as the regular output.

In monostable mode, the 4047B can be triggered by a positive or negative transition (depending on which of two input pins is used). It ignores steady input states and will also ignore additional trigger pulses that occur during the output pulse. However, a retrigger pin is provided to extend the output pulse if desired.

Complementary output pins are provided, one being active-high while the other is active-low.

To time very long periods, the 4047B was designed to facilitate connection with an external counter.

The power supply for the 4047B can be as low as 3VDC. Its maximum source or sink output current is only 1mA when powered at 5VDC, but up to 6.8mA when powered at 15VDC.

The chip is still available from manufacturers such as Texas Instruments (which markets it as the CD4047B) in through-hole and surface-mount formats. However, despite its versatility, the 4047B is less popular than dual monostable timers, described in the next section.

Dual Monostable Timers

Various timers that run only in monostable mode are available in dual format (i.e., two timers in one chip). This format became popular partly because two monostable timers can trigger each other to create an astable output, in which the pulse width, and the gap between pulses, can be set by a separate resistor-capacitor pair on each timer. This allows greater flexibility than is available when using a 555 timer.

Most dual monostable timer chips are edge-triggered by a change in input voltage, and will ignore a steady DC voltage. Consequently, the output from one timer can be connected directly to the input of another, and no coupling capacitors are necessary.

As in the 4047B, the user has a choice of two input pins for each timer, one triggered by a transition from low to high, the other triggered by a transition from high to low. Similarly, each timer has two outputs, one shifting from low to high at the start of the output cycle, the other shifting from high to low.

The values of a single resistor and capacitor determine the pulse duration of each timer.

Dual monostable timers often have the numeric sequence 4528 or 4538 in their part numbers. Examples include the HEF4528B from NXP, the M74HC4538 from STMicroelectronics, and the MC14538B from On Semiconductor. The 74123 numeric sequence identifies chips that have a very similar specification, with chip-family identifiers such as HC or LS inserted, as in the 74HC123 and 74LS123, and additional letters added as prefix or suffix. The pinouts of almost all these chips are identical, as shown in Figure 9-7. However, Texas Instruments uses its own numbering system, and datasheets should always be consulted for verification before any connections are made.

Many chips of this type are described as "retriggerable," meaning that if an additional trigger pulse is applied to the input before an output pulse has ended, the current output pulse will be extended in duration. Check datasheets carefully to determine whether a chip is "retriggerable" or will ignore new inputs during the output pulse.

The 74HC221 dual monostable vibrator (pictured in Figure 9-8) functions very similarly to the components cited above, but has slightly different pinouts.

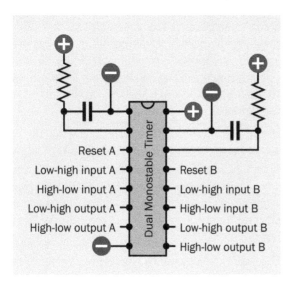

Figure 9-7. *Pin functions for most 4528, 4538, and 74123 series of dual monostable timer chips. An RC network is shown connected for each timer. Note that Texas Instruments uses different pinouts on its versions.*

Figure 9-8. *A dual-timer chip containing two monostable multivibrators that can function in astable mode if they are connected externally to trigger each other.*

Values

555 Timer Values

The original bipolar version of the 555 timer was designed to operate using a wide range of positive supply voltages, from 4.5VDC to 16VDC.

CMOS versions vary in their recommended V+ values, and datasheets must be consulted for verification.

The output of a bipolar 555 is rated to source or sink up to 200mA. In practice, the maximum current will be lower when the timer is powered at the low end of its range, around 5VDC. Attempting to source more than 50mA will pull down the voltage internally, affecting operation of the timer.

CMOS versions all impose restrictions on output current, allowing higher values for sinking than sourcing. Again, datasheets must be consulted for the values, which vary widely from one component to another.

The voltage measured on the output pin, when it is used for sourcing current, will always be lower than the power supply voltage, and a 1.7V drop is commonly specified for bipolar versions. In practice, the voltage drop that is actually measured may be less, and will vary according to the load on the output.

The voltage drop does not increase significantly with a higher supply voltage, and because it is a relatively constant value, it becomes less significant when a higher value for V+ is used.

CMOS versions of the 555 timer achieve a claimed output source voltage that is only 0.2V less than the power supply.

When choosing values for R1 and R2, a minimum for each resistor is 5K, although 10K is preferred. Lower values will increase power consumption, and may also allow overload of the internal electronics when the chip sinks current from C1. A typical maximum value for each resistor is 10M.

A high-value capacitor may cause the timer to function less accurately and predictably, because large capacitors generally allow more *leakage*. This means that the capacitor will be losing charge at the same time that it is being charged through R1 + R2. If these resistors have high values, and the capacitor has a value of 100µF or more, the rate of charge may be so low that it is comparable with the rate of leakage. For this reason, a 555 timer is not a good choice for timing intervals much greater than a minute. If a large-value capacitor is used, tantalum is preferable to electrolytic.

The minimum practical value for a timing capacitor is around 100pF. Below this, performance may not be reliable.

Although some CMOS versions may enable fast switching, the shortest practical output pulse for a 555 timer is around 10 microseconds. On the input pin, a triggering pulse of at least 1 microsecond should be used.

Time Calculation in Monostable Mode

If R1 is measured in kilohms and C1 is measured in microfarads, the pulse duration, T, in seconds, of a 555 timer running in monostable mode can be found from this simple formula:

```
T = 0.0011 * R1 * C1
```

This relationship is the same in all versions of the 555. Figure 9-9 provides a quick and convenient way to find the pulse value using some common values for R1 and C1. Resistors can be obtained with tolerances below plus-or-minus 1%, but capacitors are often rated with an accuracy of only plus-or-minus 20%. This will limit the accuracy of the pulse values shown in the chart.

Time Calculation in Astable Mode

If R1 and R2 are measured in kilohms and C1 is measured in microfarads, the frequency of pulses, F (measured in Hz) of a 555 timer running in astable mode can be found from this simple formula:

```
F = 1440 / ( ( R1 + (2 * R2) ) * C1)
```

This relationship is found in all versions of the 555. Figure 9-10 shows the frequency for common values of R2 and C1, assuming that the value of R1 is 10K. In Figure 9-11, a value of 100K is assumed for R1.

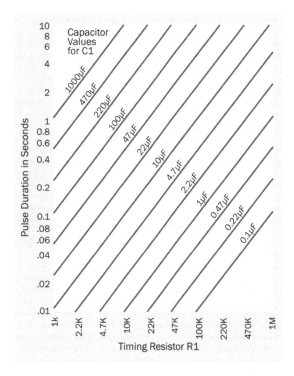

Figure 9-9. *To determine the pulse duration of a 555 timer running in monostable mode: find the value of R1 on the horizontal scale, follow its vertical grid line upward to the intersection with a green line which corresponds with the value of capacitor C1, and read across to the vertical scale providing the duration in seconds. Both axes are logarithmic.*

Dual Monostable Timers

Dual chips such as the HEF4528B from NXP, the M74HC4538 from STMicroelectronics, the MC14538B from On Semiconductor, and the 74HC123 from Texas Instruments have widely varying requirements for power supply. Some accept a limited range from 3VDC to 6VDC, while others tolerate a range of 3VDC to 20VDC. When powered with 5VDC their required input and output states are compatible with those of 5V logic chips.

Output pins of these chips source and sink no more than 25mA (much less in some instances). Because there are so many variants, they cannot be summarized here, and datasheets must be consulted for details.

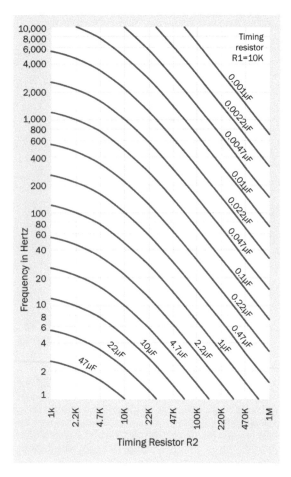

Figure 9-10. *To obtain the frequency of a 555 timer running in astable mode, when R1 has a value of 10K: find the value of R2 on the horizontal scale, follow its vertical grid line upward to the intersection with a green curve which corresponds with the value of capacitor C1, and read across to the vertical scale providing the frequency in Hertz. Both axes are logarithmic.*

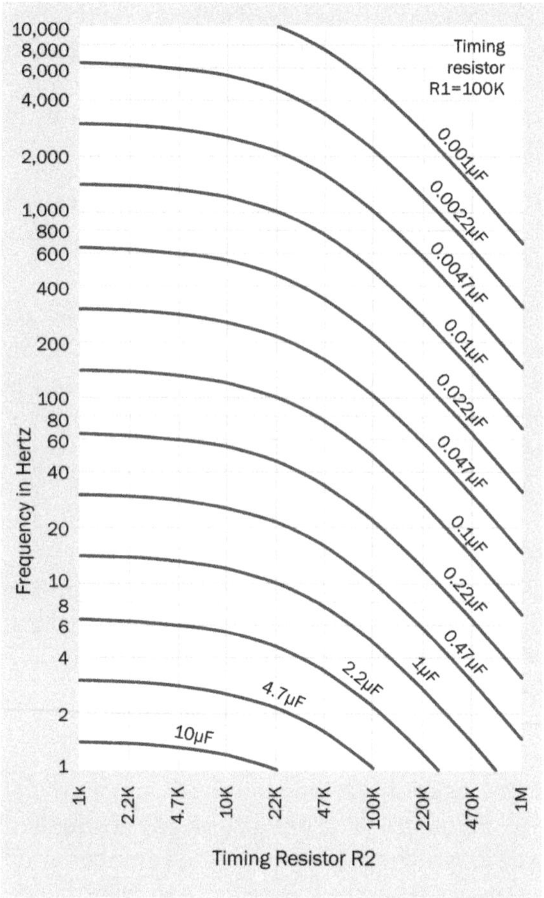

Figure 9-11. *To obtain the frequency of a 555 timer running in astable mode, when R1 has a value of 100K: find the value of R2 on the horizontal scale, follow its vertical grid line upward to the intersection with a green curve which corresponds with the value of capacitor C1, and read across to the vertical scale providing the frequency in Hertz. Both axes are logarithmic.*

As these timers are all monostable, and each timer uses just one resistor and one capacitor, the only formula required is to give the pulse time as a function of these two variables. If R is the resistor value in ohms, and F is the capacitor value in farads, and K is a constant supplied by the manufacturer, the pulse time T, in seconds, is found from the formula:

$$T = R * F * K$$

K ranges between 0.3 and 0.7 depending on the manufacturer and also on the voltage being used. Its value should be found in the manufacturer's datasheet. If R is measured in megohms and F is measured in microfarads, the formula is still valid, as the multipliers cancel each other out.

Generally speaking, these dual monostable CMOS timers are not intended for pulse duration exceeding 1 minute.

The timing capacitor should be no larger than 10μF, as it discharges directly and rapidly through the chip.

How to Use It

Where a timer is required to drive a load such as a relay coil or small motor directly, the original TTL version of the 555 timer will be the only choice. Even in this instance, a protection diode must be used across the inductive device.

For smaller loads and applications in chip-to-chip circuits, CMOS versions of the 555, including the 7555, use less power, cause less electrical interference, and are pin-compatible while using the same formulae to calculate frequency in astable mode or pulse duration in monostable mode. They are of course more vulnerable to static discharge, and care must be taken to make a connection to every pin (the capacitor that grounds the Control pin, if Control is not going to be used, is mandatory).

In dual monostable timers, unused rising-edge trigger inputs should be tied to V+ while unused falling-edge trigger inputs should be tied to ground. A Reset pin that will not be used should be tied to V+, unless that entire timer section of the chip will be unused, in which case the pin should be grounded.

To measure durations longer than a few minutes, a timer which incorporates a programmable counter to divide the clock frequency is the sensible choice. See the description of the 5555 timer that was included earlier in this entry.

The original bipolar version of the 555 remains a robust choice in hobby applications such as robotics, and its design allows some versatile variations which may even be used in logic circuits. A variety of configurations are shown in the schematics below.

555 Monostable Mode

The basic schematic for a 555 timer running in monostable mode is shown in Figure 9-12. In this particular example, a pushbutton that is liable to suffer from *switch bounce* is connected to the Input pin of the timer, which responds to the very first connection made by the pushbutton and ignores the subsequent "bounces," thus producing a "clean" output. To avoid retriggering, which results in a prolonged output pulse, the timer's output should exceed the time for which the button is likely to be pressed. The output should also exceed the duration of any possible switch bounce, which can otherwise create multiple output pulses. In the schematic, an LED is attached to the timer output for demonstration purposes.

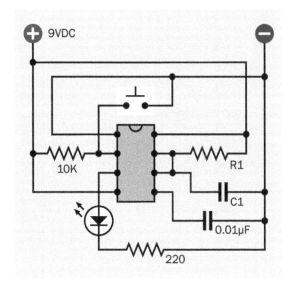

Figure 9-12. The basic monostable configuration of a 555 timer. This particular circuit debounces an input from a pushbutton switch and converts it to a clean pulse of fixed duration, powering an LED for demonstration purposes.

This circuit is shown on a breadboard in Figure 9-13. The red and blue wires, at the top of the photograph, supply 9VDC to the board. R1 is 1M, while C1 is 1μF, creating a pulse of just over 1 second. A tactile switch, just above the timer, provides the input.

Figure 9-13. *The basic monostable configuration for a 555 timer, mounted on a breadboard.*

555 Astable Mode

A basic schematic for a 555 timer running in astable mode is shown in Figure 9-14. Once again, an LED is attached to the output for demonstration purposes. If the pulse rate exceeds the persistence of vision, a small loudspeaker can be used instead, in series with a 47Ω resistor and a 100µF capacitor.

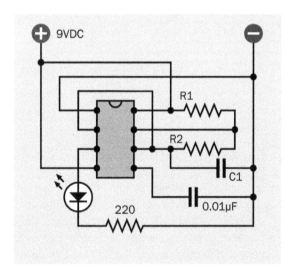

Figure 9-14. *A 555 timer with external connections and components causing it to run in astable (free-running) mode.*

Separate Control of High and Low Output Times

In Figure 9-15, a bypass diode has been added around R2. The capacitor now charges primarily through R1, as the diode has a much lower effective resistance than R2. It discharges only through R2, as the diode blocks current in that direction. Consequently, the length of the high output pulse can be adjusted with the value of R1 only, while the length of the low output pulse can be adjusted with the value of R2 only. The duration of the high output can be lower than, or equal to, the duration of the low output, which is not possible with the basic configuration of components in Figure 9-14.

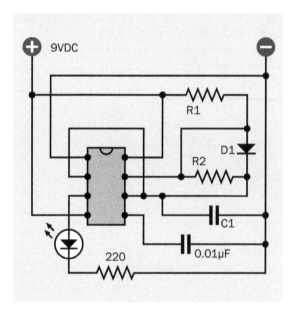

Figure 9-15. *In this circuit, a diode bypasses R2, so that the "on" time and the "off" time of the 555 timer can be set independently of each other, with R1 and R2, respectively.*

555 Fifty Percent Astable Duty Cycle: 1

In Figure 9-16, the circuit enables a fixed astable output duration of approximately 50% high and 50% low. Initially, C1 has no charge, pulling the Input of the timer low, and causing it to begin a cycle with a high pulse from the Output pin, as usual. In this demonstration circuit, the output illuminates an LED. At the same time, resistor R1 is attached to the output and charges C1. When the voltage on C1 reaches 2/3 of V+, this is communicated to the timer Input pin, which ends the "high" cycle and initiates low status on the Output pin. This starts to sink the charge from C1, through R1. When the voltage drops to 1/3 V+, this initiates a new cycle. Because only one resistor is used to charge and discharge the capacitor, we may imagine that the charge and discharge times should be identical. However, a higher load on the output will probably pull down the output voltage to some extent, lengthening the charge time. Conversely, a load on the Output pin that has low resistance will probably sink at least some of the charge from the capacitor, shortening the discharge cycle.

charges only through R2. However, in this configuration the capacitor is discharging into a voltage divider created by the two resistors. Empirical adjustment of the resistor values may be necessary before the duty cycle is precisely 50%.

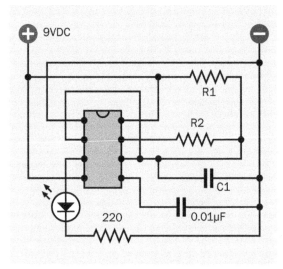

Figure 9-17. *An alternative configuration to provide an approximate 50-50 on-off duty cycle in a 555 timer.*

Use of the 555 Control Pin

In Figure 9-18, a potentiometer and two series resistors allow a varying voltage to be applied to the Control Pin. This will lengthen or reduce both the charge and the discharge times of the timing capacitor. If values for the capacitor and its associated resistors are chosen to create a frequency of approximately 700Hz, a 10K potentiometer should demonstrate more than an octave of audible tones through the loudspeaker. Other components can be substituted for a potentiometer, creating possibilities for producing pulse-width modulation. Alternatively, if a large capacitor is added between the Control pin and ground while a second 555 timer, running slowly in astable mode, applies its output to the Control pin, the charging and discharging of the capacitor will apply a smoothly rising and falling voltage. If the first 555 timer is running at an audio frequency, the output will have a "wailing siren" effect.

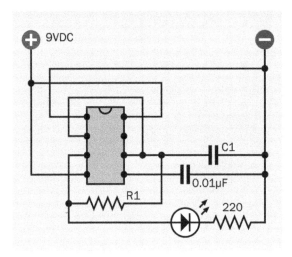

Figure 9-16. *This configuration provides an approximate 50-50 on-off duty cycle at the output pin, although the precise duration will depend on the load.*

555 Fifty Percent Astable Duty Cycle: 2

In Figure 9-17, a small modification of the basic astable circuit shown in Figure 9-14 provides another way to enable a 50% duty cycle. Compare the two schematics, and you will see that just the connection between R1 and R2 has been altered so that C1 now charges only through R1, and dis-

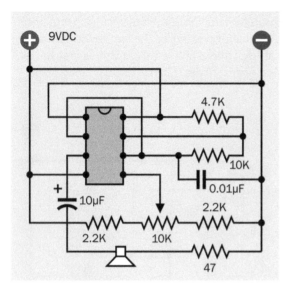

Figure 9-18. *A circuit that allows adjustment of the astable 555 frequency by increasing or lowering the voltage on the Control pin.*

Figure 9-19 shows the components specified in Figure 9-18 mounted on a breadboard.

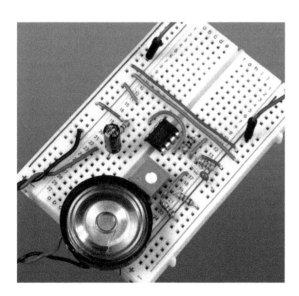

Figure 9-19. *The components in the previous schematic are shown here mounted on a breadboard. They will generate an audio output ranging between approximately 425Hz and 1,075Hz. A lower value for the timing capacitor will shift the audio range higher in frequency.*

555 Flip-Flop Emulation

The flip-flop inside a 555 timer can be accessed to control the timer's outputs. In Figure 9-20, pushbutton switch S1 applies a negative pulse to the Input pin, creating a high output from the timer, which illuminates LED D1. Normally the pulse length would be limited by the charge time of a capacitor attached to the Threshold pin, but in this circuit there is no capacitor, and the Threshold pin is hard-wired to negative ground. So, it never rises to 2/3 of positive power, and the output from the timer remains high indefinitely.

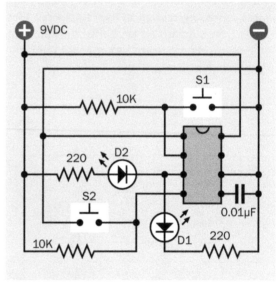

Figure 9-20. *A 555 timer can have its timing features disabled so that it functions as a flip-flop.*

However, if S2 is pressed, it grounds the Reset pin of the timer, which ends the high output and pulls the Output pin down to a low state. D1 goes out and D2 lights up, as the timer is now sinking current through it. When S2 is released, the timer output remains low and D2 remains illuminated, because the Input pin is held high by a pullup resistor. Therefore, the timer now functions in bistable mode, as a flip-flop. While this may be seen as an inappropriate use of the chip, because its full functionality is being disabled, its ability to deliver substantial current and to tolerate a wide range of supply voltages may make it more convenient to use than a digital flip-flop. See Chapter 11 for more information about flip-flops.

A 555 timer emulating a flip-flop is shown on a breadboard in Figure 9-21.

Figure 9-21. *The schematic in which a 555 timer acts as a flip-flop is shown here adapted for a breadboard.*

555 Hysteresis

The comparators inside a 555 timer enable the chip to produce hysteresis. In Figure 9-22, the Input pin and the Threshold pin are shorted together, and C1, the timing capacitor, is omitted. A 10K potentiometer, wired as a voltage divider, delivers a voltage to the Input pin ranging smoothly from V+ to negative ground. As the input dips below 1/3 V+, the Output pin goes high, lighting LED D1. Now if the input voltage gradually rises, the output remains high, even as the input rises above 1/3 V+. The output state is "sticky" because the timer does not end an output pulse until the Threshold pin tells it to, by reaching 2/3 V+. When this finally occurs, the Output pin goes low, D1 goes out, and D2 comes on, sinking current into the Output pin.

Suppose, now, the input voltage starts to go down again. Once again the output state is "sticky" because it remains low until the Input pin drops below the 1/3 level. When that happens,

the output finally flips back to a high state, D2 goes out, and D1 comes on.

In the "dead zone" between 1/3 and 2/3 of supply voltage, the timer remains in its current mode, waiting for the input to stray outside of those limits. This behavior is known as *hysteresis*, and is of special importance when processing a varying signal, such as the voltage from a temperature sensor, to control an on/off device such as a thermostat. In fact the 10K potentiometer in this demo could be replaced with a thermistor or a phototransistor, wired in series with a resistor to create a voltage divider which will have an input range compatible with the 555 timer. The hysteresis can then be adjusted by varying the supply voltage that powers the timer, as this will change the values of 1/3 V+ and 2/3 V+. Alternatively, varying the voltage on the Control pin will affect the hysteresis.

A **comparator** can provide much more versatile control of hysteresis by using positive feedback (see Chapter 6 for additional details). But the 555 timer provides a quick-and-simple substitute, and its greater ability to source or sink current enables it to be connected with a wider range of other components.

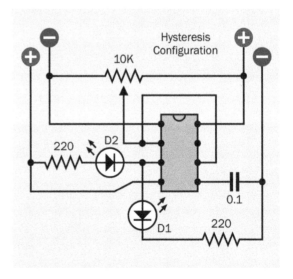

Figure 9-22. *A 555 timer wired so that it creates hysteresis, when supplied by a variable input voltage.*

555 and Coupling Capacitors

As previously noted, when a basic bipolar 555 timer (and some of its variants) is wired in mono-stable mode, it will retrigger itself indefinitely if its input remains low. One way to avoid this is by using a coupling capacitor. This will pass a transition from high to low, but will then block a steady subsequent voltage. In Figure 9-23, a phototransistor in series with a resistor provides a variable voltage to the noninverting input of a comparator. The reference voltage of the comparator is adjusted with a potentiometer, and resistor R3 provides positive feedback, ensuring that the output from the comparator will be quick and clean. The output from the 555 timer goes through a transistor to the relay, shown at the bottom.

It is important to see the function of the coupling capacitor, C3, with the pullup resistor, R2, which holds the Input pin of the 555 timer high by default. When the output from the comparator drops from high to low, C3 passes this transition to the Input pin of the timer, momentarily overcoming the positive potential, and triggering the timer. After the timer responds, however, C3 blocks any continuation of low voltage from the comparator. Pullup resistor R2 resumes its function of holding the input high, and prevents the timer from being retriggered.

555 Loudspeaker Connection

A small 8-ohm loudspeaker can be driven from the output of a bipolar 555 timer running in astable mode, but should be isolated from it with a 10µF to 100µF capacitor. A series resistor of 47Ω (minimum) should be used. See Figure 9-24.

Burst Mode

It is sometimes useful to create a short beep of fixed length in response to a button-press. The beep should terminate even if the button is held down. This "burst mode" can be achieved with the circuit in Figure 9-25, where the button connects power to a bipolar-type 555 timer running in astable mode, and an RC network applies a

decreasing potential to a 47µF capacitor, which is wired to the Reset pin of the timer. The resistor in series with the capacitor will vary the length of the beep. When voltage to the pin drops below approximately 0.3V, output from the timer stops and cannot restart until the button is released.

A resistor of greater than 1.5K may not allow the input value at the reset pin to fall below the voltage, which is necessary to enable a reset. If a lower power supply voltage than 9VDC is used, the resistor value should be higher—for example, a 5VDC power supply works well with a 1.5K to 2K resistor.

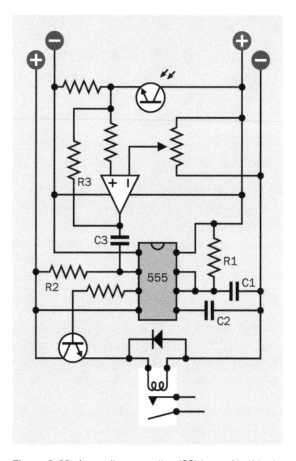

Figure 9-23. *A coupling capacitor (C3) is used in this circuit to isolate the 555 timer from a sustained low input from the comparator. The capacitor only passes a transition from high to low. The rest of the time, the pullup resistor (R3) holds the input high.*

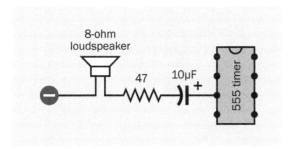

Figure 9-24. *A small 8-ohm loudspeaker can be attached through a capacitor and a resistor to the output of a bipolar 555 timer.*

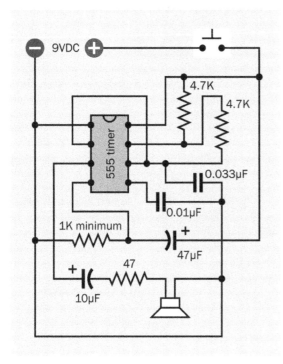

Figure 9-25. *An RC circuit, wired to apply a decreasing voltage on the Reset pin of a bipolar 555 timer, will shut off the timer shortly after it is powered up. This can be used to create a fixed-length beep in response to a button press of any duration.*

Figure 9-26 shows the components installed on a breadboard.

"You Lose" Game Sound

A timer is a simple, cheap way to create a variety of simple game sounds. The schematic in Figure 9-27 makes a groaning sound as the 100μF

capacitor wired to the Control pin of a bipolar-type 555 gradually charges through the 1K resistor. Note that if a larger resistor is paired with a smaller capacitor, the effect will differ. The 150K resistor is included to discharge the capacitor reasonably quickly in time for the next cycle.

Figure 9-26. *The "burst mode" circuit installed on a breadboard with a miniature loudspeaker.*

What Can Go Wrong

Dead Timer

Like any chip, the 555 can be damaged by over-voltage, excessive source current or sinking current, static electricity, incorrectly applied polarity of power supply, and other forms of abuse. The TTL version of the timer is fairly robust, but the CMOS type much less so. Check for obvious errors such as lack of supply voltage, incorrect or ambiguous input voltages, and unusual current draw (too high, or none at all, at the V+ pin). Use the meter probes on the actual pins of the chip, in case there is a break in the wiring that feeds them. Because timer chips are cheap, a reserve supply of them should be maintained.

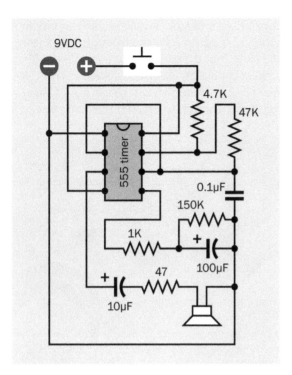

Figure 9-27. *An RC circuit, wired to apply an increasingly positive voltage to the Control pin of a bipolar timer running in astable mode, will gradually pull down the frequency at the Output pin, creating a sound that may be useful in simple game applications.*

CMOS Confused with Bipolar

The part numbers of some bipolar chips are very similar to those of some CMOS versions, and the chips look physically identical. But the CMOS version is easily overloaded, as it may source only 10mA to 20mA maximum while the TTL version is capable of 200mA. Make sure that your chips are carefully labeled when they are stored.

The Pulse that Never Ends

If a 555 timer responds correctly to a high-to-low transition on the input pin, but the output pulse continues indefinitely, check the voltage on pin 6 to see if the timing capacitor is charging above 2/3 of V+. While a 555 can run from 5VDC, a high-current device on the Output pin can pull down the voltage inside the chip to the point where the capacitor never charges sufficiently to end the cycle.

Also check that the input transition from high to low lasts for a shorter time than the pulse. A persistent low input can retrigger the timer.

Erratic Chip Behavior

Possible causes include:

- Floating pins. The Input pin, in particular, should always be connected with a defined voltage (via a 10K pullup resistor, if necessary), and must not be allowed to float at an indeterminate potential.

- Voltage spikes. A timer can be triggered by transients from other components, especially inductive loads. If the input to a monostable timer dips for even a fraction of a second, the timer will initiate a new cycle. A *protection diode* should be used in conjunction with an inductive load.

- Voltage spikes can also introduce variations in the pulse train from an astable timer.

- TTL versions of the 555 timer will tolerate a wide range of supply voltage, but if a voltage regulator is not used, fluctuations in voltage can have unpredictable consequences.

Interference with Other Components

Because the bipolar version of a 555 timer creates a voltage spike when its output changes state, it can interfere with the normal function of other components, especially CMOS chips. A 0.1μF bypass capacitor can be applied between the timer's V+ pin and ground.

Erratic Behavior of Output Devices

If a 555 timer powers an output device such as a relay, and the relay is not opening or closing in a reliable manner, first check that it is receiving sufficient voltage. If the 555 timer is powered with 5VDC, its output will be only around 4VDC.

This problem can be avoided by using the output from the timer to control the voltage on the base of a transistor which switches a separate source of power to the relay coil.

Fatal Damage Caused by Inductive Loads

While it is possible to drive an inductive load such as a small motor or relay directly from a TTL 555 timer, two precautions should be taken. First, the motor or the coil of the relay should have a clamping diode added around it, as is standard practice. Second, because the output of the timer is capable of sinking current as well as sourcing current, it can be protected from sinking back-EMF by inserting a diode in series with the load. This is illustrated in Figure 9-28.

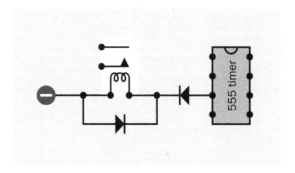

Figure 9-28. *In addition to a standard protection diode clamped around an inductive load such as a relay coil, the 555 timer can be protected from back-EMF by adding a diode in series. The series diode must of course be rated to carry sufficient current through the coil. When choosing a relay, allowance must be made for the voltage drop that will be imposed by the series diode.*

logic gate

10

Only basic **logic gates** are included in this entry—that is, components that perform a Boolean logic operation on two to eight inputs (or one input, in the case of an inverter) to create a single high or low logical output.

OTHER RELATED COMPONENTS

- **flip-flop** (see Chapter 11)

What It Does

A **logic gate** is a circuit that delivers an output, either high or low, depending on the states of its two inputs, either or both of which can be high or low.

Some gates may have more than two inputs, and an *inverter* only has one input, but the basic gates all conform with the two-input, one-output model. The components that constitute a logic gate are almost always etched into a wafer inside a silicon chip.

In a digital computer, a *high logic state* is traditionally close to 5VDC and represents a value of 1 in *binary arithmetic*, while a *low logic state* is traditionally close to 0VDC and represents a binary 0. Modern devices may use a lower voltage for the high state, but the principle is still the same.

A small network of logic gates can perform binary addition, and all other operations in a digital computer are built upon this foundation.

Origins

The concept of digital logic originated in 1894, when English mathematician George Boole announced his invention of a form of algebra (now referred to as *Boolean algebra*) to analyze com-

binations of two logical states that could be interpreted as "true" and "false." This concept had few practical applications until the 1930s, when Claude Shannon saw that because a basic switch has two states, Boolean algebra could enable analysis of complex networks of switches that were being used in telephone systems.

Because the state of a switch could also be used to represent the values 0 or 1 in binary arithmetic, and because a transistor could function as a switch, Boolean algebra was implemented in solid-state digital computing equipment.

How It Works

While conventional arithmetic uses *arithmetical operators* to represent procedures such as addition or multiplication, Boolean algebra uses *Boolean operators*. The operators of special interest in digital electronics are named AND, NAND, OR, NOR, XOR, and XNOR.

Although each gate actually contains multiple transistors, it is represented by a single logic symbol, as shown in Figure 10-1. The names of the Boolean operators are customarily printed all in caps. A gate requires a power supply and a connection with negative ground, separate from its inputs, but these connections are omitted from gate schematics because they are assumed to exist.

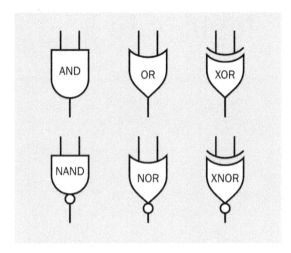

Figure 10-1. *Six types of two-input logic gates are used in digital electronics, although the XNOR gate is rare, as it has few applications. The names are customarily printed in uppercase letters.*

The functions of the gates with two inputs can be defined in electrical terms. In Figure 10-2, the four possible combinations of inputs are tabulated in the left column, with red indicating a high input and black indicating a low input. The corresponding output from each gate is shown beneath its name. This kind of tabulation is known as a *truth table*, as it is derived from Boolean algebra which originally concerned itself with "true" and "false" states.

Input states of a two-input gate	Gate Outputs					
	AND	NAND	OR	NOR	XOR	XNOR
● ●	●	●	●	●	●	●
● ●	●	●	●	●	●	●
● ●	●	●	●	●	●	●
● ●	●	●	●	●	●	●

Figure 10-2. *The four possible combinations of input states in a 2-input logic gate are shown at left. The corresponding output from each gate is shown beneath its name. Red indicates a high state, while black indicates a low state.*

The truth table assumes that *positive logic* is being used. Negative logic is very uncommon, but if it were used, the red dots in the truth table would correspond with low inputs and outputs, while the black dots would correspond with high inputs and outputs.

Inversion

The small circles appended to the outputs of NAND, NOR, and XNOR gates mean that the output of each gate is inverted compared with the AND, OR, and XOR gates. This can be seen by inspection of the output states shown in Figure 10-2. The circles are known as *bubbles*.

Sometimes logic symbols are shown with a bubble applied to one input, as in Figure 10-3. In these cases, the circle indicates that an input must be inverted. More than one gate may be needed to achieve this logic function in an actual circuit. The style is often used to show the inner workings of an IC, using a minimum number of logic symbols.

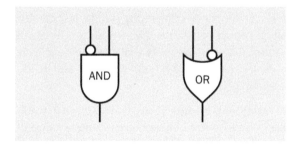

Figure 10-3. *The circle in a logic gate symbol indicates that a signal is being inverted. Circles can be inserted at gate inputs, but in a real circuit a separate inverter is likely to be needed to create this effect.*

Single-Input Gates

Two gates exist that have a single input and a single output, shown in Figure 10-4. The *buffer* should not be confused with the symbol for an **op-amp** or a **comparator**. (Those components always have two inputs.) The output state of a buffer is the same as its input state, but the component may be useful to deliver more current or to isolate one section of a circuit from another.

When a bubble is appended to a buffer, it becomes a NOT gate, more commonly known as an *inverter*. Its function is to create an output state that is opposite to its input state.

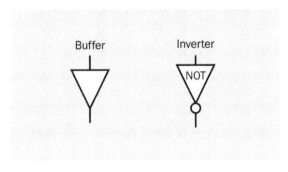

Figure 10-4. *The two logic gates that have only one input and one output. Note that in some schematics showing internal logic of ICs, the bubble on an inverter may be found on the input side instead of the output side.*

Gates with More than Two Inputs

AND, NAND, OR, and NOR gates can have any number of inputs, as suggested in Figure 10-5, although practical factors usually limit the inputs to a maximum of eight.

Input states of a gate with more than two inputs	Gate Outputs			
	AND	NAND	OR	NOR
All low	●	●	●	●
All high	●	●	●	●
At least 1 low and at least 1 high	●	●	●	●

Figure 10-5. *The previous table has been modified to show the outputs from logic gates that have more than two inputs. XOR and XNOR gates are not included in the table, because a strict interpretation of their logic requires that a unique output state exists if one input is high while the other is low.*

The rules can be summarized like this:

- Output from an AND gate: Low if any of its inputs is low, high if all of its inputs are high.
- Output from a NAND gate: High if any of its inputs is low, low if all of its inputs are high.
- Output from an OR gate: High if any of its inputs is high, low if all of its inputs are low.
- Output from a NOR gate: Low if any of its inputs is high, high if all of its inputs are low.

In the case of XOR and XNOR gates, their logic requires that a unique output state must exist if one input is high while the other input is low.

In fact, so-called three-input XOR gates do exist, an example being the 74LVC1G386 chip, in which the output is high if all three inputs are high, or if one input is high, but not if two inputs are high or no inputs are high. Further discussion of more-than-two-input XORs is outside the scope of this encyclopedia.

Boolean Notation

For reference, the original written notation for Boolean operators is shown in Figure 10-6. Unfortunately, the notation for these operators was never properly standardized, and in more than one instance, multiple symbols acquired the same meaning. The letters P and Q are often, but not always, used to represent two input states that can be true or false.

- The use of a horizontal line above a symbol, to indicate that its state has been reversed, has carried over to datasheets where this notation can show that an output state from any digital chip is inverted. The line is known as a *bar*.

Arithmetical Operations

Suppose we wish to sum two binary numbers, each containing two digits. There are four digits altogether, and depending on their values, there are 16 different possible addition sums, as shown in Figure 10-7.

If A0 and B0 represent the rightmost digits of the two numbers being added, and S0 is the sum of those two digits, inspection of the figure shows that the sum can be derived using just three rules:

1. If A0 = 0 and B0 = 0, then S = 0.

2. If A0 and B0 have opposite states, then S0 = 1.

3. If A0 = 1 and B0 = 1, then S0 = 0, and carry 1 to the next place left.

If A0 and B0 are the two inputs to an XOR logic gate, the output of the gate satisfies all three rules, except the need to carry 1 to the next place left. This last function can be satisfied with an AND gate. The function of two gates is known as a *half adder*, and is shown in the top section of Figure 10-8.

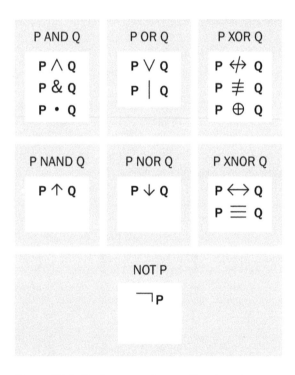

P AND Q

$P \wedge Q$
$P \& Q$
$P \cdot Q$

P OR Q

$P \vee Q$
$P \mid Q$

P XOR Q

$P \nleftrightarrow Q$
$P \neq Q$
$P \oplus Q$

P NAND Q

$P \uparrow Q$

P NOR Q

$P \downarrow Q$

P XNOR Q

$P \longleftrightarrow Q$
$P \equiv Q$

NOT P

$\neg P$

Figure 10-6. *Boolean operators as they have been expressed in written notation. Lack of standardization has resulted in more than one symbol representing some of the operators.*

When we consider the next pair of binary digits to the left, the situation now becomes more complicated, because we may be carrying 1 into this addition sum from the previous stage, and we still need to be able to to carry 1 out (if necessary) to the next stage. An assembly of five logic gates

can deal with this, and their combination is known as a *full adder*. This is shown in the bottom section of Figure 10-8.

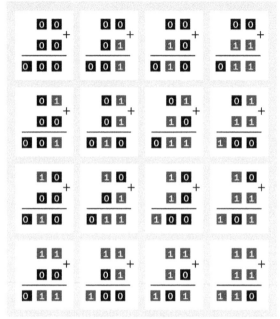

Figure 10-7. *Sixteen different addition sums are possible, when summing two binary numbers of two digits each.*

The combination of XOR and AND gates shown in Figure 10-8 is not the only one that works to add binary numbers. However, it may be the most intuitively obvious.

Other Operations

Binary arithmetic remains the most important application of logic gates, but individually packaged gates are seldom used for that purpose anymore. They were long since subsumed into large multifunction computing chips.

Single gates still have application in small systems, or to modify the inputs and outputs of microcontrollers, or to convert the output from one complex digital chip to make it compatible with the input of another. This last application is often referred to as *glue logic*.

Applications for single gates are discussed in "How to Use It" on page 103.

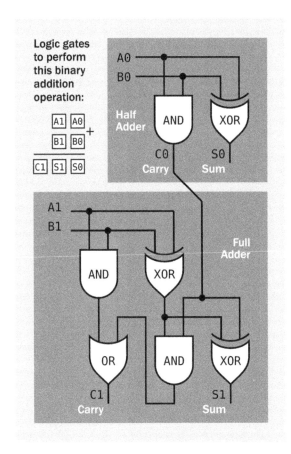

Figure 10-8. *Logic gates can be used to add binary numbers, using a high input or output to represent a binary 1 and a low input or output to represent a binary 0. This schematic shows one possible way for gates to add two two-digit binary numbers.*

Variants

Chips containing logic gates were introduced in the 1960s. The 7400 NAND chip, from Texas Instruments, was the first of a series that became so influential, the same basic part numbers (with letters added before, after, and among the digits) are still used today. An example of a currently available through-hole 7400 chip is shown in Figure 10-9.

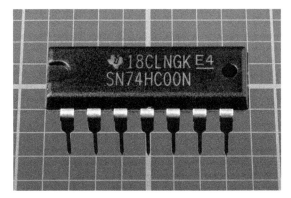

Figure 10-9. *A modern version of a 7400 chip containing four NAND gates.*

Initially, these chips conformed with a *transistor-transistor logic* (*TTL*) standard that had been invented at TRW in 1961 and introduced in commercial products by Sylvania in 1963. It established the now-familiar standard of 5VDC for the power supply. Many logic chips now use other voltages, but the term "high" still means an input or output that is near to the supply voltage, while "low" means an input or output that is near to negative ground. The exact definition of "near" will be found in datasheets for the chips.

The *7400 series* was successful partly because it was engineered for compatibility. The output from one gate could be connected directly to the input of another gate, with a few bypass capacitors added on a circuit board to suppress voltage spikes caused by rapid switching. Earlier components had not been so easy to interface with each other. The new standard dominated the industry to the point where dozens of manufacturers started making chips that conformed with it, and a single board could mix-and-match chips from multiple vendors.

Because many logic chips acquired part numbers that began with 74, they are often referred to as the 74xx series, where other digits (sometimes more than two) can be substituted for xx. This avoids ambiguity, as the very first chip in this format was a NAND gate that had 7400 as its actual part number. In the text below, 7400 will refer to that specific chip, whereas 74xx will refer to the whole series of chips.

RCA introduced a competing family of logic gates in 1968, using CMOS transistors. As each part number began with a 4 and contained four digits, this was referred to as the *4000 series*. The CMOS chips were slower and more expensive, but tolerated a wider range of power supply voltages (3V to 12V, initially). Their biggest advantage was that they used much less current. This was important, as TTL chips created a lot of waste heat. The lower power consumption of CMOS also enabled one chip to control the inputs of many others, which simplified circuit design. This one-to-many relationship is known as *fanout*.

Ultimately, CMOS chips transcended their early limitations. While they were reserved initially for battery-powered devices in which very low power consumption was more important than speed, CMOS is now used almost everywhere, still maintaining its advantage of low current (almost zero, in fact, while a chip is quiescent) while equalling the speed of TTL. However, CMOS logic chips are very often pin-compatible with the old TTL components, and modern CMOS part numbers are often derived from the old 74xx series.

Most CMOS logic chips in the old 4000 series are still available, and may be used in situations where a power supply greater than 5VDC is convenient.

Part Numbers

As the performance of semiconductors gradually improved, successive *families* of logic chips were introduced, identified by one-letter, two-letter, or three-letter acronyms. The acronym was inserted into the part number, so that a 7400 NAND gate in the HC (high-speed CMOS) family became a 74HC00 NAND gate.

Because these chips were available from multiple sources, the part number was also preceded with one or more letters indicating the manufacturer. And because each chip was manufactured in different versions (for example, some complied

with military specifications, while others didn't), letters were also appended to the end of the part number. Today, the appended letters may indicate whether the chip is of the old through-hole format, or conforms with a more recent surface-mount format.

Summing up:

- Prefix: manufacturer ID.
- Numerals, omitting any letters in the middle: Chip functionality.
- Middle letters: Chip family.
- Suffix letters: Package format.

Thus, for example, the actual part number for a 74HC00 NAND chip could be SN74HC00N, where the SN prefix indicates that it is manufactured by Texas Instruments and the N suffix means that it is in plastic dual-inline-pin (DIP) format. (The SN prefix was introduced by Texas Instruments in the earliest days of integrated circuits as an acronym for "semiconductor network," meaning that multiple transistors were "networked" on a wafer of silicon. Other manufacturers used their own schemes for part numbering, and so SN became exclusively identified with Texas Instruments.)

The system of augmenting part numbers has been further extended by inserting 1G, 2G, or 3G immediately after the family identifier, to indicate surface-mount chips that contain one, two, or three logic gates. If the "G" identifier is missing, the chip usually has four logic gates, which was the standard used in the original 74xx series. This rule applies even in surface-mount formats, where the surface-mount pads of four-gate chips have the same functions as the pinouts of the original TTL versions (except in the case of square-format surface-mount chips, which are not discussed here).

When searching a catalog to find a chip by its part number, it helps to remember that searching for a 7400 chip may not find any hits, but searching for a 74HC00 (or any other valid number containing a family identifier) is much more likely to be understood.

A key to understanding part numbers is shown in Figure 10-10. The upper part of the figure is a guide to interpreting numbers on a generic basis, while the lower part interprets the specific part number shown.

Families

As of 2013, the HC family in the 74xx series has become so widely used, it can be considered the default in the traditional DIP 14-pin format. Incremental improvements are still being made, and new families are being introduced, primarily in surface-mount formats which use lower power-supply voltages (down to around 1VDC).

Figure 10-10. *How to interpret the segments of a logic chip part number in the 74xx family (in this case, a 7400 NAND gate).*

Here is an historical summary of the most important chip families.

- 74xx: Original series of bipolar TTL chips.
- 74Hxx: Bipolar TTL, high speed, about twice as fast as the original 74xx chips, but twice the power consumption.
- 74Lxx: Bipolar TTL, lower power consumption than the original TTL, but also much lower speed.
- 74LSxx: Bipolar TTL, lower power with Schottky input stages, faster than original TTL. Some LS chips are still being manufactured.
- 74ASxx: Bipolar TTL, Advanced Schottky, intended to supercede the 74Lxx.
- 74ALSxx: Bipolar TTL, Advanced Low Power Schottky, intended to supercede the 74LSxx.
- 74Fxx: Bipolar TTL, Faster.
- 74HCxx: CMOS high-speed emulation of 74LSxx.
- 74HCTxx: CMOS but with similar high-state input voltage threshold to bipolar TTL chips, for compatibility.
- 74ACxx: Advanced CMOS.
- 74ACTxx: Advanced CMOS emulation of TTL with similar high-state input voltage threshold to TTL, for compatibility.
- 74AHCxx: Advanced Higher-Speed CMOS, three times as fast as HC.
- 74VHCxx: Very High Speed CMOS.
- 74AUCxx, 74FCxx, 74LCXxx, 74LVCxx, 74ALVCxx, 74LVQxx, 74LVXxx: Various specifications, many using power supply voltages of 3.3V or below.

In the 4000 series, an early significant improvement was the 4000B family, which allowed a higher power supply limit (18V instead of 12V) and was much less susceptible to damage by static discharge. The 4000B family almost totally replaced the old 4000 family, and most 4000B chips are still available, as they are useful in situations where a power supply delivers more than 5VDC.

- When it is referenced casually, the B at the end of a chip number in the 4000 series may be omitted. When the number is listed in a catalog, the B is included.

Chips with 45 as their first two digits were introduced as a new generation, but were not widely adopted. After that, the 4000 series ceased to evolve, and CMOS chips adopted 74xx part numbers, distinguishing themselves by the insertion of letter groups in the center of the number.

To add to the confusion, some 4000 series part numbers were appended to 74xx part numbers, so that, for example, the 74HCT4060 is designed to be compatible with the old 4060B chip.

Family Interoperability

One of the most important issues relating to chip families is their differing specifications for a low-state voltage and a high-state voltage in inputs and outputs.

The original 74xx TTL series, using a 5VDC power supply, used these approximate specifications:

- Output: 74xx voltage representing a low state (at most 0.4V to 0.5V)
- Input: 74xx input voltage interpreted as a low state (maximum 0.8V)
- Output: 74xx voltage representing a high state (at least 2.4V to 2.7V)
- Input: 74xx input voltage interpreted as a high state (minimum 2V)

This provided a safe margin of error of at least 0.4V when chips were communicating with each other.

In the CMOS 4000 family, however, logic chips required a minimum input of 3V to 3.5V to be interpreted as a high state. The minimum acceptable output from a TTL chip was below this level, creating problems if anyone should try to use the output from a TTL chip to communicate with an input on a CMOS chip.

One solution is to add a 4.7K pullup resistor to the TTL output, guaranteeing that it won't fall too low. But this wastes power, and the need for the resistor is easily forgotten. Another option is to use the HCT or ACT family of CMOS logic. The "T" in these family names indicates that they have been engineered to share the input standards of the old TTL chips. They still deliver the same high output as other types of CMOS, making them seem to be the best possible solution. Unfortunately, it does entail a compromise: the "T" chips are more sensitive to noise, among other factors.

- Ideally, chip families should not be intermingled.

Gates per Chip

Each of the original 74xx chips contained multiple gates within the limits of a uniform 14-pin through-hole format. The gates that were most commonly used had two inputs, and there were four of these gates per chip.

However, the desire for miniaturization, and the use of automatic chip-placement and soldering equipment, made one-gate and two-gate logic chips desirable and practical in surface-mount format. (Three-gate surface-mount chips exist, but are sufficiently unusual that they are not described in this encyclopedia.)

Two Inputs, Single Gate

Where a chip contains just one logic gate, it is almost always a surface-mount component, and the part number has 1G in the middle to indicate "one gate." Pad functions are shown in Figure 10-11. The layout is standardized for all logic gates, with the exception of XNOR gates, which are not manufactured in surface-mount format.

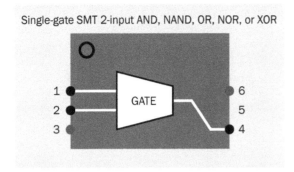

Single-gate SMT 2-input AND, NAND, OR, NOR, or XOR

Figure 10-11. *Internal configuration and solder-pad functions for a two-input surface-mount single-gate logic chip that can contain an AND, NAND, OR, NOR, or XOR gate. XNOR gates are not manufactured in this format.*

In the figure, a gate is shown in generic form, indicating that it may be an AND, NAND, OR, NOR, or XOR gate, depending on the part number of the chip. Inputs are on the left of the gate, while its output is on the right. The chip does not have a solder pad in position 5, but the pad at top right is identified as pin 6 for consistency with the numbering pattern in other surface-mount components where six pads are common.

The generic part numbers for single-gate surface-mount two-input logic chips are shown here, with letter x indicating that letter sequences are likely to be inserted to indicate manufacturer, logic family, and chip format:

- AND gate: x74x1G08x
- OR gate: x74x1G32x
- NAND gate: x74x1G00x
- NOR gate: x74x1G02x
- XOR gate: x74x1G86x

Three Inputs, Single Gate

AND, NAND, OR, and NOR single gates are available with more than two inputs. Their output is determined by rules shown in Figure 10-5. XOR and XNOR gates are not included in the table, because a strict interpretation of their logic requires that a unique output state exists if one input is high while the other is low.

The pad functions for a surface-mount single-gate logic chip with three inputs are shown in Figure 10-12. The generic part numbers for these chips are shown below. Again, each x indicates that letter sequences are likely to be inserted to indicate manufacturer, logic family, and chip format.

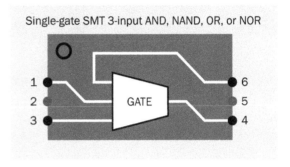

Single-gate SMT 3-input AND, NAND, OR, or NOR

Figure 10-12. *Internal configuration and solder-pad functions for a three-input surface-mount single-gate logic chip that can contain an AND, NAND, OR, or NOR gate.*

The generic part numbers for single-gate surface-mount three-input logic chips are shown here, with each x indicating that letter sequences are likely to be inserted to indicate manufacturer, logic family, and chip format:

- AND: x74x1G11x
- NAND: x74x1G10x
- OR: x74x1G32x
- NOR: x74x1G27x

Single Gate, Selectable Function

A few surface-mount chips can emulate a variety of two-input gates, by using appropriate external connections. The internal logic of one example, with generic part number x74x1G97x (an actual example would be Texas Instruments SN74LVC1G97), is shown in Figure 10-13. Depending which pin is grounded and which other pins are used as inputs, the chip can emulate all five of the most commonly used gates. To achieve this, however, some inputs have to be inverted.

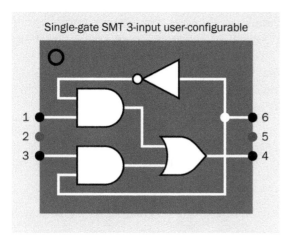

Figure 10-13. *Internal configuration for a configurable surface-mount chip that can emulate various two-input logic gates, depending which inputs are used and which are grounded. Some inputs have to be inverted to emulate some gates.*

Two Inputs, Dual Gate

Two-input surface-mount AND, NAND, OR, NOR, and XOR gates are available in dual layout (two gates per chip). The internal logic and pad functions are shown in Figure 10-14. The generic part numbers for these chips are shown here. Again, each x indicates that letter sequences are likely to be inserted to indicate manufacturer, logic family, and chip format.

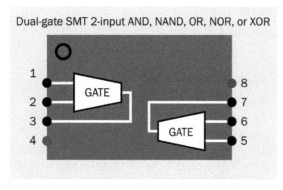

Figure 10-14. *Internal configuration and solder-pad functions for a two-input surface-mount dual-gate logic chip that can contain two AND, NAND, OR, NOR, or XOR gates. XNOR chips are not manufactured in this format.*

The generic part numbers for dual-gate surface-mount two-input logic chips are shown below, with each x indicating that letter sequences are likely to be inserted to indicate manufacturer, logic family, and chip format:

- AND: x74x2G08x
- NAND: x74x2G00x
- OR: x74x2G32x
- NOR: x74x2G02x
- XOR: x74x2G86x

Original 74xx 14-Pin Format

Each of the original 74xx TTL chips contained multiple gates within the limits of a uniform 14-pin chip format. The available options were, and still are:

- Quad 2-input: Four gates with two inputs each
- Triple 3-input: Three gates with three inputs each
- Dual 4-input: Two gates with four inputs each
- Dual 5-input: Two gates with five inputs each
- Single 8-input: One gate with eight inputs

The five-input chips have become so uncommon that they are not described in this encyclopedia.

Quad Two-Input 74xx Pinouts

14-pin DIP 74xx quad two-input logic chips are available in AND, NAND, NOR, XOR, or XNOR versions, all of which have an internal layout shown in Figure 10-15. The layout is unchanged in surface-mount format. The gates are shown in generic form, as the layout remains the same regardless of which type of gate is in the chip. All the gates in any one chip are of the same type. The four connections leading to a gate are its inputs, while the single connection from a gate is its output.

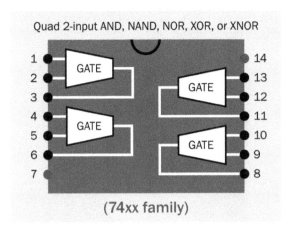

Figure 10-15. *In a 14-pin quad two-input 74xx logic chip, the AND, NAND, NOR, XOR, and XNOR versions all share this generic layout.*

- The 14-pin quad two-input OR chip has different pinouts from all the other 74xx logic chips. It is shown in Figure 10-16.

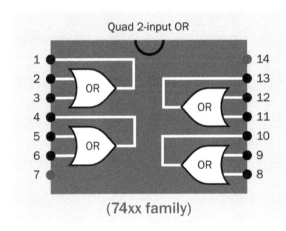

Figure 10-16. *In a quad two-input 74xx OR chip, this layout is used, which is different from that used in all the other quad two-input logic gates.*

Triple Three-Input 74xx Pinouts

The AND, NAND, and NOR versions of a 14-pin DIP 74xx triple three-input logic chip all have an internal layout shown in Figure 10-17. The layout is unchanged in surface-mount format. The gates are shown in generic form, as the layout remains the same regardless of which type of gate is in the chip. All the gates in any one chip

are of the same type. Three connections leading to a gate are its inputs, while a single connection from a gate is its output.

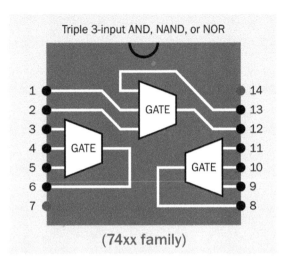

Figure 10-17. *In a 14-pin triple three-input 74xx logic chip, the AND, NAND, and NOR versions all share this generic layout.*

- The 14-pin triple three-input OR chip has different pinouts from all the other 74xx logic chips. It is shown in Figure 10-18.

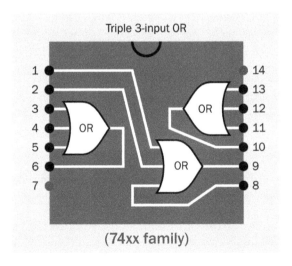

Figure 10-18. *In a triple three-input 74xx OR chip, this layout is used, which is different from that used for all the other triple three-input logic gates.*

Dual Four-Input 74xx Pinouts

A 14-pin DIP 74xx dual four-input logic chip contains two four-input gates. The AND, NAND, and NOR versions all have an internal layout shown in Figure 10-19. The layout is unchanged in surface-mount format. The gates are shown in generic form, as the layout remains the same regardless of which type of gate is in the chip. All the gates in any one chip are of the same type.

- There is no OR chip of the 14-pin dual four-input type.

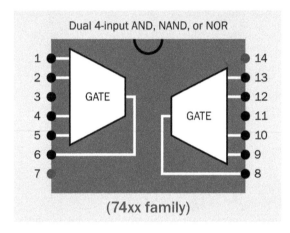

Figure 10-19. *In a 14-pin dual four-input 74xx logic chip, the AND, NAND, and NOR versions all share this generic layout. There is no 74xx OR chip with four inputs per gate.*

Single Eight-Input 74xx Pinouts

A 14-pin DIP 74xx single eight-input NAND chip contains one eight-input gate, as shown in Figure 10-20. The layout is unchanged in surface-mount format.

- There is no AND chip of the 14-pin single eight-input type.

A 14-pin eight-input logic chip in the 74xx series, able to function as both an OR and a NOR, is shown in Figure 10-21. The output from the NOR gate is connected with pin 13, but also passes through an inverter to create an OR output at pin 1. (Because a NOR gate is equivalent to an inverted-OR, when its output is inverted again, it returns to being an OR.)

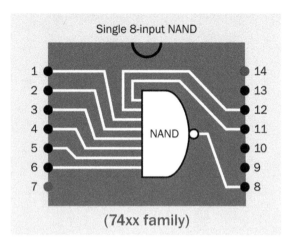

Figure 10-20. *The internal layout of single eight-input NAND chip in the 14-pin 74xx series. There is no 74xx AND chip with eight inputs per gate.*

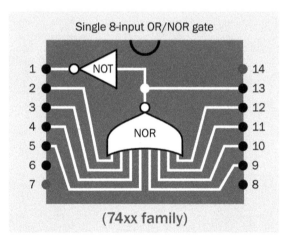

Figure 10-21. *The internal layout of single eight-input OR/NOR chip in the 14-pin 74xx series. Pin 13 has the NOR output, while pin 1 has the OR output.*

The following list shows the generic part numbers for DIP and surface-mount versions of 14-pin logic chips in the 74xx series that have two or more inputs per gate. As before, an x indicates that letter sequences are likely to be inserted to indicate manufacturer, logic family, and chip format.

- Quad 2-input AND: x74x08x
- Quad 2-input NAND: x74x00x
- Quad 2-input OR: x74x32x
- Quad 2-input NOR: x74x02x
- Quad 2-input XOR: x74x86x
- Quad 2-input XNOR: x74x266x
- Triple 3-input AND: x74x11x
- Triple 3-input NAND: x74x10x
- Triple 3-input OR: x74x4075x
- Triple 3-input NOR: x74x27x
- Dual 4-input AND: x74x21x
- Dual 4-input NAND: x74x20x
- Dual 4-input NOR: x74x4002x
- Single 8-input NAND: x74x30x
- Single 8-input OR/NOR: x74x4078x

74xx Inverters

Single, dual, and triple inverter packages in the 74xx series are available in surface-mount format only. Their internal arrangement is shown in Figures 10-22, 10-23, and 10-24.

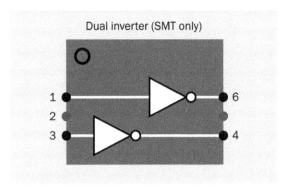

Figure 10-23. *The internal layout of a 74xx series logic chip containing two inverters. This is available in surface-mount format only.*

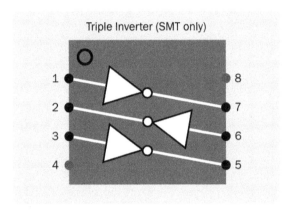

Figure 10-24. *The internal layout of a 74xx series logic chip containing three inverters. This is available in surface-mount format only.*

In the 14-pin format, a hex inverter chip (containing six inverters) is available, as shown in Figure 10-25. The layout is the same for DIP and surface-mount formats.

Generic part numbers for inverter chips are as follows:

- Single inverter: x74x1G04x
- Dual inverter: x74x2G04x
- Triple inverter: x74x3G14x
- Hex inverter: x74x04x

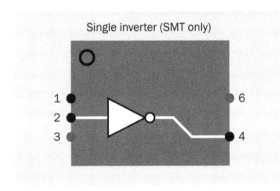

Figure 10-22. *The internal layout of a 74xx series logic chip containing one inverter. This is available in surface-mount format only. Pin 5 is absent. Pin 1 is not connected.*

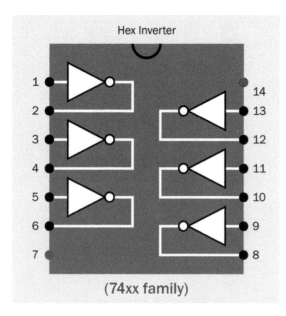

Figure 10-25. *The internal layout of a 14-pin 74xx hex inverter logic chip, containing six inverters. This layout is the same for DIP and surface-mount versions.*

Additional Variations

Some chips in the 74xx series (both DIP and surface mount versions) have variants with open drain or open collector outputs, while others have inputs that are configured as Schmitt triggers. These variants will be found as hits when searching supplier websites for logic chips by gate name and number of inputs.

Pinouts in the Original 4000 Series

Each of the original 4000 CMOS chips contained multiple gates within the limits of a uniform 14-pin chip format. The available options were, and still are:

- Quad 2-input: Four gates of two inputs each
- Triple 3-input: Three gates of three inputs each
- Dual 4-input: Two gates of four inputs each
- Single 8-input: One gate of eight inputs

In the 4000 family, 14-pin quad two-input logic chips are available in AND, OR, NAND, NOR, XOR,

or XNOR versions, all of which have an internal layout shown in Figure 10-26. The gates are shown in generic form, as the layout remains the same regardless of which type of gate is in the chip. All the gates in any one chip are of the same type. The four connections leading to a gate are its inputs, while the single connection from a gate is its output.

Unlike the 74xx family, the quad two-input OR chip in the 4000 family has the same pinouts as the other types of quad two-input logic chips.

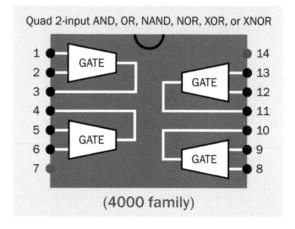

Figure 10-26. *In the 4000 family, the AND, OR, NAND, NOR, XOR, and XNOR versions of a quad two-input logic chip all share this generic layout.*

In the 4000 family, a 14-pin triple three-input logic chip contains three three-input gates. The AND, OR, NAND, and NOR versions all have an internal layout shown in Figure 10-27. The gates are shown in generic form, as the layout remains the same regardless of which type of gate is in the chip. All the gates in any one chip are of the same type. The three connections leading to a gate are its inputs, while a single connection from a gate is its output.

Unlike the 74xx family, the triple three-input OR chip in the 4000 family has the same pinouts as the other types of triple three-input logic chips.

In the 4000 family, a 14-pin dual four-input logic chip contains two four-input gates. The AND, NAND, OR, and NOR versions all have an internal layout shown in Figure 10-28. The gates are shown in generic form, as the layout remains the same regardless of which type of gate is in the chip. All the gates in any one chip are of the same type. Each pair of connections leading to a gate are its inputs, while the single connection from a gate is its output.

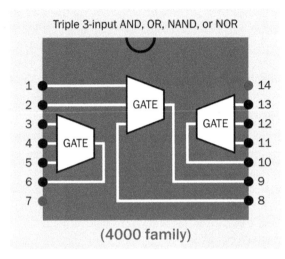

Figure 10-27. *In the 4000 family, the AND, OR, NAND, and NOR versions of a triple three-input logic chip all share this generic layout.*

Note that the 4000 family does have a dual four-input OR chip, whereas the 74xx family does not.

In the 4000 family, a 14-pin eight-input logic chip with AND and NAND outputs is available, as shown in Figure 10-29.

The following list shows the generic part numbers for 14-pin logic chips in the 4000 family that have two or more inputs per gate (in actual part numbers, letters will be substituted where an x appears):

- Quad 2-input AND: x4081x
- Quad 2-input NAND: x4011x
- Quad 2-input OR: x4071x
- Quad 2-input NOR: x4001x
- Quad 2-input XOR: x4070x
- Quad 2-input XNOR: x4077x

- Triple 3-input AND: x4073x
- Triple 3-input NAND: x4023x
- Triple 3-input OR: x4075x
- Triple 3-input NOR: x4025x
- Dual 4-input AND: x4082x
- Dual 4-input NAND: x4012x
- Dual 4-input OR: x4072x
- Dual 4-input NOR: x4002x
- Single 8-input AND/NAND: x4068x

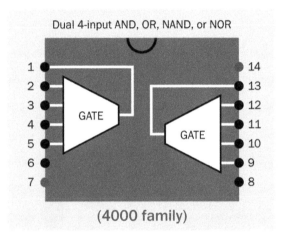

Figure 10-28. *In the 4000 family, the AND, OR, NAND, and NOR versions of a dual four-input logic chip all share this generic layout.*

4000 Series Inverters

In the 4000 family, the 4069B is a 14-pin hex inverter chip (containing six inverters), as shown in Figure 10-30. This has the same pinouts as the x74x04x chip.

How to Use It

Which Family

In DIP format, the HC family has existed for more than 30 years, and has become established as a widely used default choice.

In surface-mount formats, the choice of family will largely be determined by the choice of supply voltage.

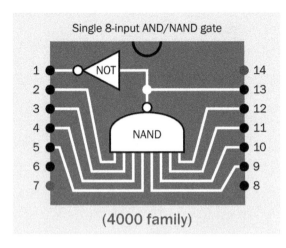

Figure 10-29. *In the 4000 family, a single eight-input AND/NAND chip has this internal layout. The inverted output from the NAND gate becomes an AND output from pin 1 of the chip.*

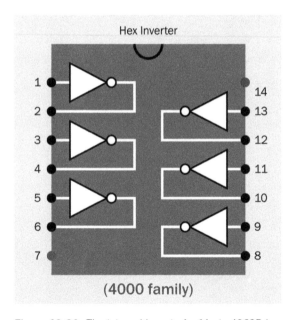

Figure 10-30. *The internal layout of a 14-pin 4069B hex inverter logic chip, containing six inverters. This layout is the same as for the x74x04x chip.*

Although the 4000 series is now more than 40 years old, it may still be useful where a 5VDC power supply is not required for other reasons in a circuit and would be added purely to power a 74xx series logic gate or other digital chip. If a

circuit contains a 9VDC or 12VDC relay, for instance, a Darlington pair may be used with that voltage to drive the relay, and an old-school 4000 series logic chip could share the same supply. The relay coil would need a clamping diode to protect the logic chip from transients.

Applications

The output from a logic chip may be used as an input for a microcontroller, to enable multiple inputs to share one pin. An eight-input NAND gate, for instance, could combine the inputs from eight normally on motion sensors. If just one sensor responds to an intrusion, the gate output would change from high to low.

Logic gates may be useful in any simple device that has to respond to a single, specific combination of inputs. A digital combination lock is one example; games of chance are another. Most simple dice simulations use logic gates to convert the output from a counter to drive a dice-pattern of LEDs.

A logic gate may be used as an interface between an electromechanical **switch** and a circuit containing digital chips. A 10K pullup or pulldown resistor prevents the gate input from floating when the switch is open. A buffer can be used for this purpose, or an inverter, or any "spare" gate on a logic chip that is already in the circuit. One input of the chip can be tied to the positive power supply or negative ground, to create an appropriate input from the chip when the switch, attached to the other input, is opened or closed.

A jam-type **flip-flop** can be used to *debounce* a switch input. See Chapter 11 for details. If two NOR or two NAND gates are unused in a circuit, they can form a flip-flop.

In the original CMOS 4000 family, a positive output may be capable of driving an LED if the current does not exceed 5mA with a power supply of 5VDC or 10mA with a power supply of 10VDC. Note that the output voltage will be pulled down significantly by these loads. In the 74HCxx family, chips can source or sink as much as 20mA, but here again the output voltage will be pulled down. Note that the total limit for all outputs from a 74HCxx chip is around 70mA.

The output from a logic chip can be passed through a buffer such as the 7407, which has an open-collector output capable of sinking as much as 200mA. This enables direct drive of modest loads, so long as they are not inductive.

Solid-state relays and **optocouplers** can be driven directly from logic chips, as they draw very little current. A solid-state relay can switch 50A or more.

What Can Go Wrong

Two problems are common when using CMOS digital chips: damage from *static electricity*, and erratic behavior caused by *floating pins*.

Static

The early 4000 series CMOS chips were especially vulnerable, but more recent CMOS designs generally include diodes at the inputs, which reduces the risk. Still, logic chips should be protected by inserting them into anti-static foam or enclosing them in conductive wrappers until they are installed in a board. While handling chips, it is good practice to be grounded, ideally using a wrist-mounted ground wire.

Floating Pins

Any pin which is unconnected in a logic chip is considered to be "floating," and can pick up signals by capacitive coupling, possibly disrupting the behavior of the chip and also causing power consumption, as the ambiguous pin state will tend to prevent that gate in the chip from entering quiescent mode.

Generally speaking, input pins in a TTL logic chip that are not being used for any purpose should be tied to the positive voltage supply, while unused CMOS pins should be tied to negative ground.

Family Incompatibilities

As previously noted, older TTL logic chips may deliver a "high" output voltage that is lower than the minimum expected by newer CMOS logic chips. The best option is not to mix families, but if chips are stored carelessly, some intermingling can occur. Part numbers should be checked if one chip appears to be ignoring output from another.

Overloaded Outputs

If a circuit calls for a logic chip with an open-collector output, and a regular chip is used by mistake, it will almost certainly be damaged.

Output Pulled Down

If the output from one logic chip is connected with the input of another logic chip, and if the output from the first chip is also connected to an **LED**, the LED may pull down the output voltage so that the second chip will not recognize it as a high state. As a general rule, a logic output can drive an LED, or can drive another logic chip, but not both. Very-low-current LEDs, which draw as little as 2mA, may be acceptable.

Incorrect Polarity and Voltages

Logic chips can be knocked out by applying incorrect polarity, or voltage to the wrong pin, or the wrong voltage. Modern logic chips tolerate a very limited voltage range, and a 74xx series chip will be irrevocably damaged if it is used where a 4000 series chip was specified for a power supply higher than 6VDC.

If a chip is inserted upside-down, it will probably be damaged when voltage is applied.

Bent Pins

Like all through-hole chips, DIP logic chips can be inserted accidentally with one or more pins bent underneath the chip. This error is very easy to miss. The bent pins will not make contact with any socket that is used, and the chip will behave unpredictably. Check for proper pin insertion with a magnifying glass if necessary.

Unclean Input

Logic chips expect a clean input without voltage spikes. A 555 timer of TTL type generates spikes in its output which can be misinterpreted as multiple pulses by the input of a logic chip. A CMOS-type 555 timer is more suitable for connection with logic chips.

If a pushbutton, rotational encoder, or electro-mechanical switch provides a high or low input, the input must be debounced. In hardware, this is traditionally done with a **flip-flop**. It can also be done with code in a microcontroller.

Analog Input

The input of a logic chip can be connected directly with a thermistor, phototransistor, or similar analog component, but only if there is some certainty that the voltage at the input pin will remain within the range that is acceptable to the chip. In the case of a phototransistor, for example, it should be exposed to a limited, known range of light intensity.

In general, it is best to avoid applying intermediate-voltage signals to a digital logic input, as they can create unpredictable output, or output of an intermediate voltage. A **comparator** can be placed between the analog source and the digital logic chip, or a logic chip with a Schmitt-trigger input can be used.

flip-flop | 11

The term **flip-flop** is sometimes printed with a space instead of a hyphen, but the hyphenated form seems to predominate in the United States. Therefore, the hyphen is included here. The term *flipflop* (with neither a hyphen nor a space) is sometimes seen, but is unusual. The acronym *FF* is confined mostly to logic diagrams or schematics.

The term *latch* is sometimes used interchangeably with *flip-flop* but is assumed here to describe a minimal asynchronous circuit that responds immediately and transparently to an input. A flip-flop can function as a latch and also as a synchronous device which is opaque, meaning that the input does not flow directly through to the output.

OTHER RELATED COMPONENTS

- **counter** (see Chapter 13)
- **shift register** (see Chapter 12)

What It Does

Transistors enable logic gates; logic gates enable flip-flops; and flip-flops enable many mathematical, storage, and retrieval functions in digital computing. Most flip-flops today are embedded in much larger integrated circuits that have complex functions. However, they are still available as separate components in chip form, and will be discussed on that basis here.

A flip-flop is the smallest possible unit of memory. It can store a single bit of data, represented by either a high or low *logic state*. (A full explanation of logic states is included in the **logic gate** entry. See Chapter 10.) Flip-flops are especially useful in **counters**, **shift registers**, and *serial-to-parallel converters*.

A flip-flop circuit can be classified as a form of *bistable multivibrator*, as each of its outputs is stable in one of two states until an external trigger stimulates it to "flip" from one state and "flop" into the other. (For a comprehensive discussion of monostable and astable multivibrators, see the **timer** entry in Chapter 9.)

An *asynchronous* flip-flop will respond immediately to a change of input, and can be used for applications such as debouncing the signal from an electromechanical **switch** or building a *ripple* **counter**. More often, a flip-flop is *synchronous*, meaning that a change in input state will be unrecognized until it is enabled by a low-to-high or high-to-low transition in a stream of pulses from an external *clock*.

How It Works

Every flip-flop has two outputs, each of which may have a high or low state. When the flip-flop is functioning normally, the outputs will be in opposite logical states, one being high while the other is low. These outputs are typically identified as Q and NOT-Q (the latter term meaning a letter Q with bar printed above it, sometimes referred to verbally as "Q-bar"). In datasheets and other documents where a bar symbol cannot be represented easily above a letter, the NOT-Q output may be represented as letter Q with an apostrophe after it, as in Q'.

Almost always, in a schematic diagram, a flip-flop is represented by a simple rectangle, with inputs and outputs identified by letters and other marks. Because a description of the inner workings is necessary before the different types of flip-flop can be understood, schematic symbols for various flip-flops will not be introduced until "Variants" on page 116.

The simplest flip-flop contains two logic gates whose function can be most easily understood if the inputs are controlled by a SPDT switch. It can be created from two NAND gates or two NOR gates, as described next. This type of component can be described as:

- *asynchronous*: Will accept data on an impromptu basis, as it is not synchronized with a clock.

- *jam-type*: Colloquial equivalent of *asynchronous*. The input is jammed in at any time, forcing an immediate change of output.

- *transparent*: The input state flows straight through to the output.

NAND-Based SR Flip-Flop

Figure 11-1 shows two NAND gates attached to a SPDT switch, with two pullup resistors. When either of the NANDs has a floating input from the switch, the pullup resistor attached to that input will maintain it in a high state. The data inputs for the NAND gates are labeled S and R, meaning Set and Reset, giving this component its name as an *SR* flip-flop:

- In a NAND-based SR flip-flop, a low state is considered an active logic input, as indicated by the bar placed above each letter.

- A high state is considered an inactive logic output.

The schematic style in this figure, with diagonally crossing conductors, is universally used and easily recognizable. The equivalent schematic in Figure 11-2, which might be created by circuit-drawing software, has the same functionality but would not be immediately recognizable as a flip-flop. The "classic" crossed-conductor representation is preferable.

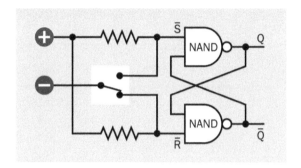

Figure 11-1. *The schematic for a simple NAND-based SR-type flip-flop, with a switch and pullup resistors added to control two inputs.*

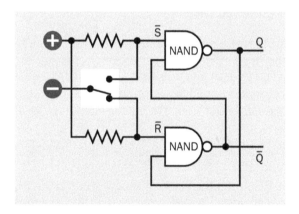

Figure 11-2. *An alternative component layout for an SR flip-flop, functionally identical to the previous schematic, but not so easily recognizable. The layout with a pair of diagonally crossing conductors has become so ubiquitous, it should be considered to be a standard.*

The first step toward understanding the behavior of flip-flops is to recall the relationship between the two inputs and the output of NAND or NOR gates. This is shown in Figure 11-3, where red indicates a high logic state and black indicates a low logic state.

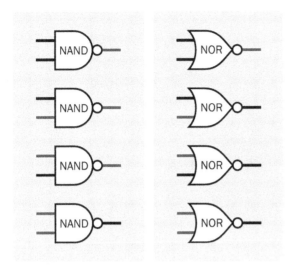

Figure 11-3. *The four possible input combinations for a NAND gate and a NOR gate, with the corresponding logical outputs. A flip-flop can be built from two NANDs or two NORs.*

The behavior of a NAND gate can be summarized:

- Both inputs high: Output low.
- Other input combinations: Output high.

Figure 11-4 shows a series of snapshots of the SR flip-flop circuit as the switch moves from one position, through an intermediate state where it makes no connection, to the other position. Remember that in this circuit, the active logical input state is low, and the active logical output state is high.

In the top panel, the pullup resistor of the lower NAND is overwhelmed by the direct connection to negative ground, which holds the R input in a low state. The other input of this gate is irrelevant, because the output from a NAND will be high if either of its inputs is low. So, the lower NAND has a high output, which feeds back to the secondary input of the upper NAND gate. The S input of this gate is high, because of the pullup resistor. Because both inputs of this gate are now high, its output is low, which feeds back to the lower gate. The lower gate doesn't change its

output, because either of its low inputs is enough to keep its output high. So, the circuit is in equilibrium. A high state on the NOT-Q output is known as the *Reset* state for a NAND-based flip-flop.

The second panel shows what happens if the switch now moves up into a neutral, disconnected position. The R input of the lower NAND now becomes high, because of the pullup resistor. But this NAND still has one low input, supplied by the output of the upper gate, so its output remains high, and the circuit is still in equilibrium. This is known as the *Hold* state for the NAND circuit.

Suppose the switch bounces to and fro between the states shown in the first two panels. The output from the circuit won't change. This shows that the circuit provides a method for eliminating *switch bounce*—the very fast, momentary spikes that occur when the mechanical contacts of a switch open and close.

The third panel shows what happens if the switch now moves to its upper position. The top input of the upper NAND gate is now pulled low. Consequently, its output goes high. This feeds back to the lower gate. Its other input is high because of the pullup resistor. With both of its inputs high, its output goes low. The gate outputs have flipped and swapped values. A high state on the Q output is known as the *Set* state for a NAND-based flip-flop.

The circuit still remains in equilibrium even if the switch returns to its central, disconnected position shown in the bottom panel. Therefore, the debouncing capability of the circuit works equally well for both positions of the switch.

NOR-Based SR Flip-Flop

Figure 11-5 shows a similar circuit using two NOR gates attached to an SPDT switch. Because the NOR gates function differently, this circuit uses active-high input logic, and pulldown resistors are needed instead of pullup resistors. The output from the circuit still uses active-high logic, and is identical with the NAND-based circuit in this respect, although the relative positions of the Q and NOT-Q outputs have been swapped.

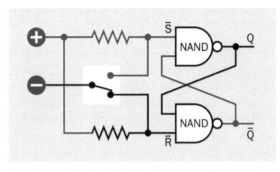

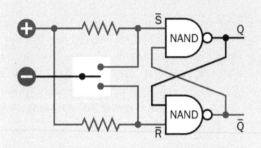

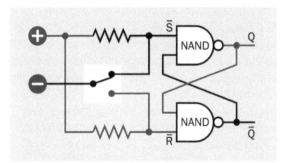

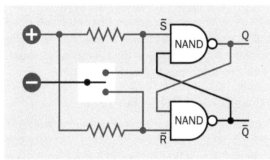

Figure 11-4. *Four snapshots of transitions in a NAND-based SR flip-flop as the switch moves down and up through an intermediate no-connection zone. See text for details.*

- In a NOR-based SR flip-flop, a high state is considered an active logic input, as indicated by the absence of a bar placed above the letters S and R.
- A high state is considered an active logic output.

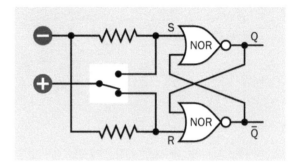

Figure 11-5. *The schematic for a simple SR flip-flop using NOR gates instead of NANDs.*

In the case of the NOR gate:

- Both inputs low: Output high.
- Other input combinations: Output low.

Figure 11-6 shows a series of snapshots as the switch moves from one position to the other, through intermediate states where it makes no connection. Remember that in this circuit, the active logical state is high at both the inputs and the outputs.

In this circuit, as in the previous circuit using NAND gates, it will ignore switch bounce, allowing the gate outputs to remain stable.

Forbidden States

Either of the circuits described so far depicts an SR flip-flop, regardless of whether it is NAND-based or NOR-based. Its input and output states are summarized in Figure 11-7. However, as this table suggests, there are some input states that create problems.

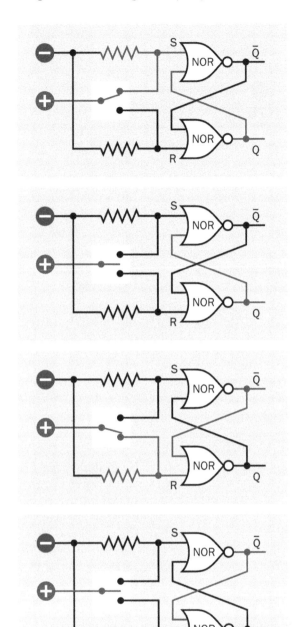

Figure 11-6. *Four snapshots of a NOR-based SR flip-flop, showing the consequence of changing switch positions, comparable with the NAND-based flip-flop.*

Flip-Flop Inputs		Flip-Flop Outputs			
		NAND-based		NOR-based	
S	R	Q	Q̄	Q	Q̄
●	●	● Problem ●		Same as Previous	
●	●	●	●	●	●
●	●	●	●	●	●
●	●	Same as Previous		● Problem ●	

Figure 11-7. *A table of input states and the consequent output states for NAND-based and NOR-based SR flip-flops.*

In either the NAND-based flip-flop or the NOR-based flip-flop, the output when the switch is in its unconnected center position will remain the same as when the switch was in its previous position. This is the usefulness of the flip-flop: it remembers the previous state. These situations are identified as "Same as Previous" in the table.

The pullup resistors (in a NAND-based flip-flop) and pulldown resistors (in a NOR-based flip-flop) are intended to guarantee that both inputs will be high (NAND) or both inputs will be low (NOR) even when the switch makes no connection. Therefore, it should be impossible for both inputs to be low (NAND) or high (NOR).

But what happens if the circuit is powered up with the switch in the unconnected position? One input of each gate is controlled by the output of the other gate. But what will those outputs be?

In the NAND-based version, the outputs from the NANDs will be low while the chips are powering up. As soon as the NAND chips are functioning, each of them will sense that it has one input high and one input low, so it will change its output to high.

But now that each chip has a high output, it will feed back to the secondary input of the other chip. Now both chips have both inputs high. This will cause them both to change their outputs to low—but this will feed back again, flipping the outputs back to high again. In fact, if the gates are absolutely identical, the circuit will oscillate very rapidly. This is sometimes known as *ringing*.

In real life, the gates will not be absolutely identical, and eventually one of them will respond fractionally ahead of the other, tipping the circuit into the state shown either in the second panel or the fourth panel in Figure 11-4. But which chip will win? There is no way of knowing. This is known as a *race condition*, and the winner is unpredictable.

A similar but opposite situation occurs in the NOR-based flip-flop if it is powered up with the switch in the disconnected position, and the S and R outputs are both low, because of the pulldown resistors. Here again it will be a race condition.

We can address the problem by making a rule that the switch must always be in one position or the other when the flip-flop is powered up. But what if there is a faulty switch? Or what if a power interruption occurs while the switch is changing position?

Another problem occurs if the switch makes one contact a fraction before it breaks the other contact. This would result in both S and R inputs being low, in a NAND flip-flop. The same state could occur if a separate logic circuit is driving the S and R inputs, and an error causes it to make S and R both low. This is shown in Figure 11-8. Because the output from a NAND gate is always high if at least one of its inputs is low, both gates now have a high output, and the circuit is stable.

The problem is, the states of the outputs from a flip-flop should always be opposite to each other. If both of them are high, this can create logic problems in the rest of the circuit attached to the flip-flop.

- In a NAND-based SR flip-flop, if S and R are both low, this is known as a *forbidden state* or a *restricted combination*.

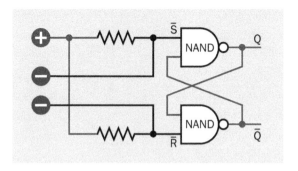

Figure 11-8. *What happens when both S and R inputs to a NAND flip-flop are low as a result of an error in a separate control circuit.*

A similar problem afflicts a NOR-based SR flip-flop, except that the forbidden state will occur when the S and R inputs are both high.

- In a NOR-based SR flip-flop, if S and R are both high, this is a *forbidden state* or a *restricted combination*.

The SR flip-flop is useful as a switch debouncer, but for computing applications, it is vulnerable to errors.

The JK Flip-Flop

Because the name of the JK flip-flop shares the initials of Jack Kilby, who won a Nobel prize for his fabrication of the world's first integrated circuit, some people speculate that this type of flip-flop was named after him. The attribution seems implausible, and may have gained currency simply because a flip-flop was the first device that Kilby happened to build when he was developing an integrated circuit.

Regardless of how it came to be named, the JK design is shown in Figure 11-9. This is commonly referred to as a JK *latch*. The electromechanical switches that were shown driving the SR flip-flop, along with pullup or pulldown resistors, are no longer included, because the inputs at positions J and K are assumed to come from other devices that have properly defined high and low states. Their behavior may be unpredictable, but neither of them will ever have a floating state.

This is a *gated* circuit, meaning that an additional input stage blocks direct access to the output stage, and it is also a *synchronous* circuit, as it uses a train of pulses at a clock input. Two three-input NAND gates are placed in front of a NAND-based SR flip-flop, and they address the problem of simultaneous identical inputs by using crossover feedback from the second stage to the first stage (via the conductors at top and bottom of the schematic).

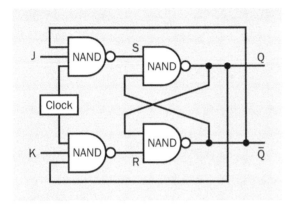

Figure 11-9. *The basic circuit for a clocked JK flip-flop, using two additional NAND gates prepended to an SR flip-flop.*

Versions of a JK flip-flop are possible using NOR gates, but are less common. Only the NAND-based version will be considered here.

In the case of a three-input NAND gate:

- All three inputs high: Output low.
- Other input combinations: Output high.

Because of the additional pair of NANDs, the circuit now recognizes a high input as logically active, instead of the low-active input in the previous SR flip-flop using NAND gates. Consequently, two simultaneous high inputs might be expected to create the type of forbidden state that was caused by two simultaneous low inputs previously. However, in Figure 11-10, the top and bottom panels show that simultaneous high inputs at J and K will support two possible valid outputs, where the state at Q is always opposite to the

state at NOT-Q. In fact, when both inputs are high, a positive pulse on the clock input will *toggle* the outputs (i.e., they will switch places). In fact, the toggling will continue so long as the clock input is high. Consequently, this type of flip-flop is intended for use with short clock pulses.

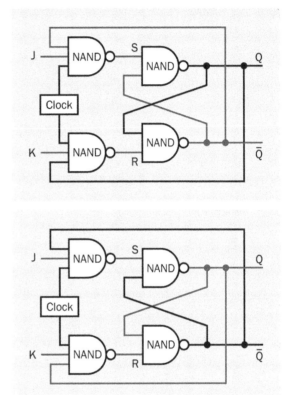

Figure 11-10. *When the J input and the K input are both high, this no longer causes a forbidden state. The combination will toggle the outputs of the flip-flop between the two states shown here.*

Master-Slave Flip-Flop

A more stable form is shown in Figure 11-11 where yet another stage has been added, this one being a "master" to the first. In fact, this configuration is known as a *master-slave* flip-flop, the slave stage being driven by the master stage but remaining inactive until a low clock input at the master stage passes through an *inverter* to become a high clock input at the slave stage. The master and slave stages thus take turns, one being activated by a high clock pulse while the other is activated by the low part of the pulse cycle. The output from the slave stage cannot feed back to the master stage while the clock pulse is still high, and thus the timing issue in the single-stage JK latch is eliminated. Because the master-slave version of the JK configuration is not *transparent*, it is correctly known as a flip-flop rather than a latch.

In addition, Preset and Clear inputs may be added to override the clock to Set or Reset the outputs. These inputs are active-low.

Figure 11-12 summarizes the behavior of a JK master-slave flip-flop that is triggered by the falling edge of each clock pulse (shown as a downward-pointing arrow in the Clock column of the table). Note that the output will be delayed while the slave stage waits for the second part of each clock cycle.

The letter X in the table indicates that the state in that cell is irrelevant.

When J and K are both low, the states of Q and NOT-Q will remain the same as in the previous cycle, and this is still referred to as a *Hold* state. When J and K are both high, the outputs toggle, meaning that their new states will be opposite to the previous states.

D-Type Flip-Flops

A D-type flip-flop places an inverter between two inputs to guarantee that they will always be in opposite states, and uses a clock signal to copy their states to a pair of logic gates.

When an inverter is added between the inputs in this way, either an SR flip-flop or a JK flip-flop can become a D-type flip-flop. Figure 11-13 shows the simplest possible D-type circuit, added to a basic SR flip-flop. Only one data input is now required (customarily labeled D), because it drives the other through the inverter.

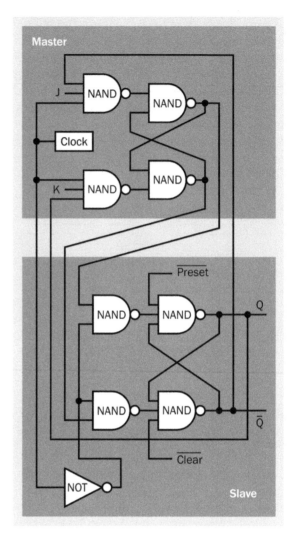

Figure 11-11. *A master-slave circuit that drives one flip-flop with another.*

Flip-Flop Inputs					Outputs	
Clock	Preset	Clear	J	K	Q	Q̄
X	●	◦	X	X	◦	●
X	◦	●	X	X	●	◦
↓	◦	◦	●	●	Same as Prev	
↓	◦	◦	●	◦	●	◦
↓	◦	◦	◦	●	◦	●
↓	◦	◦	◦	◦	Toggle	

Figure 11-12. *A table showing inputs and outputs for a JK master-slave flip-flop.*

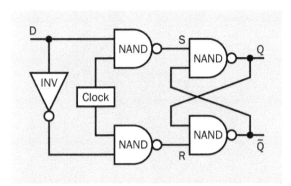

Figure 11-13. *A simple D-type flip-flop. The inverter guarantees that the state of one input will always be opposite to the state of the other.*

Figure 11-14 uses snapshots to show how the circuit responds to changing input and clock conditions. Initially, in the top panel, the data input is high, the clock input is high, and the Q output is high. In the second panel, the clock goes low, causing the output from the upper NAND gate in the input stage to change from low to high. But the upper NAND gate in the output stage still has one low input, so its state remains unchanged. In fact, the S and R inputs of the output NANDs are now both high, which creates the hold condition.

In the third panel, the D input changes from high to low, but this has no effect so long as the clock is low. The D input can fluctuate repeatedly, and nothing will happen until the clock goes high, as shown in the fourth panel. Now the clock copies the new D input state through to the output.

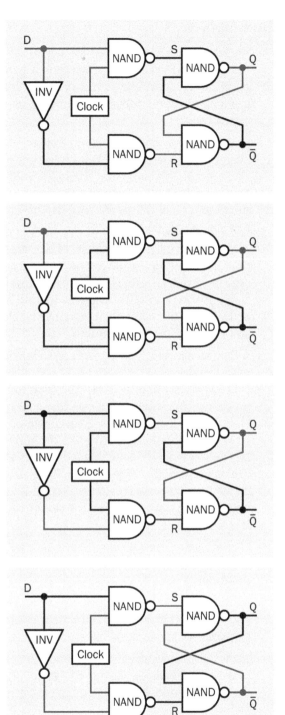

Figure 11-14. *Four snapshots showing the behavior of a D-type flip-flop.*

Summary

- An SR flip-flop can be used for switch de-bouncing, but in other applications it can enter an unacceptable race condition if its inputs and power supply are not carefully controlled.

- A JK flip-flop is gated, meaning that an SR circuit is preceded with an input stage and a clock input. This eliminates the race condition, adds the ability to toggle the outputs, but requires a very brief clock input. The circuit is edge-triggered.

- A master-slave flip-flop consists of two flip-flops, one driving the other. They can be JK type or SR type. The first flip-flop is activated by a positive clock state, while the second is activated by the subsequent negative clock state. Timing issues are resolved.

- A D-type flip-flop is gated with an inverter between the inputs, so that they cannot be simultaneously high or low. Consequently, only one input, labeled D, is needed. A high state on the D input causes a Set condition, while a low input causes a Reset condition, but only when the clock copies the status of the inputs through to the outputs. The status of the outputs remains stable (the flip-flop enters a hold condition) after the clock goes low.

- The JK circuit used to be widely used, because of its versatility. The D-type circuit now predominates.

- A T-type (toggling) flip-flop exists but is uncommon, and is not included in this encyclopedia.

- Many flip-flop circuits exist in addition to the ones that have been illustrated here. Only the most commonly cited circuits have been included.

A chip containing two positive-edge triggered D-type flip-flops is shown in Figure 11-15. Each flip-flop in this component has its own data, set, and reset input and complementary outputs.

Figure 11-15. *This chip contains two positive-edge triggered D-type flip-flops.*

Variants

A selection of schematic symbols representing flip-flops is shown in Figure 11-16. Letters S, R, J, K, or D define the type of flip-flop. Q and NOT-Q are the outputs. CLK is the clock input but may alternatively be identified with letter E, meaning *Enable*. SRCK or SCLK may also identify it, the abbreviations being intended to mean "serial clock.

A triangle preceding CLK indicates that the flip-flop is positive-edge triggered. A circle, properly termed a *bubble*, preceding the triangle, indicates that the flip-flop is negative-edge triggered. In other locations, the bubble indicates that the input (or output) is inverted; it means the same thing as a bar printed above the text abbreviation, and indicates active-low logic. Synchronous inputs are shown on the left side of the flip-flop with the CLK input, while asynchronous inputs (if any) are shown above and below the flip-flop rectangle.

Using these guidelines, the examples in Figure 11-16 can be decoded:

1. An unclocked SR flip-flop with active-low inputs (probably NAND-based).

2. An unclocked SR flip-flop with active-high inputs (probably NOR-based).

3. An SR flip-flop with active-high inputs, pulse-triggered by an active-high clock input.

4. A JK flip-flop with active-high inputs, edge-triggered by a rising-edge clock input. The bubble on the lower Q output means the same thing as a letter Q with a bar printed above it.

5. A D-type flip-flop pulse-triggered by an active-low clock input.

6. A D-type flip-flop edge-triggered by a falling-edge clock input.

7. A JK flip-flop with active-high inputs, edge-triggered by a rising-edge clock input, with asynchronous active-low Preset and Clear inputs.

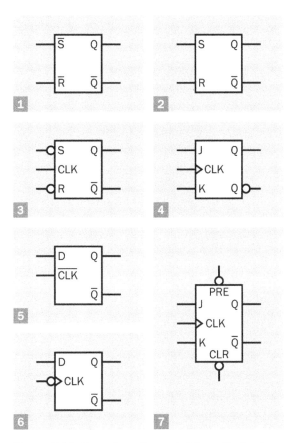

Figure 11-16. *The schematic symbol for a flip-flop is an annotated rectangle. See text for an explanation of the letters and marks.*

Packaging

Only about 10% of the flip-flops listed by a typical parts warehouse are through-hole chips. The rest are now surface-mount. Still, even if the search is narrowed further to through-hole packages in the 74xx and 4000 series, at least 100 types still exist. They provide opportunities in education and prototyping work, even though they are less often used as standalone components.

A package often contains more than one flip-flop. Dual and quad arrays are common. The flip-flops may be independently clocked, or may share a single clock input; datasheets should be checked carefully for details. Octal flip-flops, such as the D-type 74x273, are intended for use as eight-bit registers.

Many of the older flip-flops are numbered in the 74xx series of logic chips. See Chapter 10 for a detailed guide to this numbering system and the various logic families. D-type flip-flops include 74x74, 74x75, 74x174, and 74x175, where an acronym for the logic family is substituted for the x. Old-style CMOS flip-flops include the 4042B D-type latch, the 4043B quad NOR SR flip-flop, and the 4044B quad NAND SR flip-flop. The last two are synchronous, and both allow two Set inputs, labeled S1 and S2 in the datasheet.

Examples of JK flip-flops include the 74x73, 74x76, and 74x109.

Values

As is the case with other logic chips, most flip-flops in the through-hole 74xx series are intended for 5VDC power supply while the older 4000 series may tolerate up to 18VDC. Surface-mount versions may use voltages as low as 2VDC.

See "Variants" on page 93 for a discussion of acceptable high and low logic input voltages. On the output side, the 4000 series chips are able to source less than 1mA at 5VDC, but the 74HCxx series can manage around 20mA.

If a flip-flop is used for high-speed operation, the following values must be considered:

- t_S Setup time: The minimum time in nanoseconds for an input to be constant before the next clock pulse can process it.

- t_H Hold time: The minimum time in nanoseconds for an input to persist after the active edge of a clock pulse that has processed it. The interaction between a clock pulse and an input state takes a brief but measurable amount of time; errors may occur if the clock is given less than that amount of time to do its job.

- t_{CO} Clock-to-output: The elapsed time after an active clock edge, before the output changes. This is a function of the internal workings of the chip, and may be broken down into low-to-high and high-to-low output transitions, as follows.

- T_{PLH} Propagation to Low-to-High: The elapsed time after an active clock edge, before a low-to-high swing occurs at an output. This may not be identical to T_{PHL}.

- T_{PHL} Propagation to High-to-Low: The elapsed time after an active clock edge, before a high-to-low swing occurs at an output. This may not be identical to T_{PLH}.

- f_{MAX} Maximum clock frequency for reliable operation.

- $t_{W(H)}$ The minimum high clock pulse width in nanoseconds.

- $t_{W(L)}$ The minimum low clock pulse width in nanoseconds.

In a shift register or counter, where multiple flip-flops are cascaded but they share the same clock, the t_{CO} of one flip-flop must be shorter than the hold time of the next flip-flop, to allow the input of data to be complete before the window of opportunity is over.

How to Use It

The asynchronous SR flip-flop is of primary use in debouncing switches. Examples are the single MAX6816, dual MAX6817, and octal MAX6818.

D-type flip-flops are widely incorporated in frequency dividers, which are used to count pulses and display a binary output. If the NOT-Q output is wired back to the D input, the pulse stream to the clock input will have the following effect:

1. Suppose the initial D state is low and the initial state of the NOT-Q output is low.

2. The first high clock pulse propagates the low D state into the flip-flop.

3. The next low clock state forces the NOT-Q output high. This feeds back and creates a high D input.

4. The second clock pulse propagates the high D state into the flip-flop.

5. The next low clock state pulls the NOT-Q output low. This feeds back and creates a low D input.

The sequence then repeats. Only one high output is generated at NOT-Q (or at Q) for every two clock pulses; thus the circuit can become a divide-by-two **counter**. If the Q output is tapped to serve as the clock input for another flip-flop, that circuit now has a divide-by-four output. A series of many flip-flops can be chained together, so long as the propagation of signals along the chain is fast enough to occur before the next clock pulse. This is known as an *asynchronous counter*.

For more information on the use of counters, see Chapter 13.

While flip-flops have tended to be integrated with other components in digital computing, they are still used as registers where 8 or 16 bits of serial data must be assembled at a time, prior to being disseminated as parallel data.

What Can Go Wrong

Ambiguous Documentation

For reasons which are unclear, instructional texts and tutorials can be erratic when describing flip-flops:

- A truth table may fail to clarify whether the circuit uses active-high or active-low logic.
- Truth tables from different sources are often inconsistent in their representation of current and future output states, and may even fail to include the clock status in a clocked flip-flop.
- Tutorials may include logic diagrams for some types of circuit, but not others.
- NOR gates may be used, without any mention that NAND gates can also be used (and may be more common or convenient).
- The active-low or active-high status of inputs in an SR flip-flop may not be shown.

Bearing this in mind, manufacturer datasheets should be consulted whenever possible as the primary source of information.

Faulty Triggering

In many cases, a flip-flop designed for edge triggering can give erroneous results if it is level-triggered, and vice versa. Rising-edge-triggered flip-flops must be distinguished from falling-edge-triggered flip-flops. As always, it is important for similar parts that have similar functions to be stored separately.

Metastability

The behavior of flip-flops has been described in this entry under ideal conditions, where they are operating well within parameters established by the manufacturer. In reality, non-ideal scenarios may occur, especially where inputs such as data and clock, or clock and reset, are almost simultaneous. This may be difficult to avoid if a signal is received from an external source such as a sensor, with no way to control its arrival time. If the input occurs within the setup time or the hold time of a clock pulse, the flip-flop may be unable to determine whether the input precedes or follows the clock.

This may lead to *metastability*, meaning an unpredictable output and/or oscillations that take several clock cycles to settle into a stable state. If the output from a flip-flop may be used by two separate components with slightly different response times, one may interpret the oscillating output as a high state while the other interprets it as low. In a computing circuit, metastability can lead to calculation errors or a system crash. To avoid these issues, limits in datasheets should be observed. Attention should be paid to the manufacturer specifications for minimum setup time and hold time, so that the circuit has sufficient opportunity to recognize a signal and respond.

One solution to metastability is to connect multiple flip-flops in series, all sharing a common clock signal. This will tend to filter out irregularities, at the expense of requiring additional clock cycles if the flip-flops are not transparent.

Metastable-hardened flip-flops minimize metastability but cannot eliminate it completely.

Other Issues

Problems that tend to affect digital chips generally are listed in the section of the entry on logic gates (see "What Can Go Wrong" on page 105).

shift register

The term **shift register** is rarely hyphenated. In this encyclopedia, no hyphen is used.

A shift register can function as a *queue*, but this term is more usually applied to software. When the output from the last stage of a shift register is connected back to its input, it can function as a *ring counter*, but that application is described in the **counter** entry of this encyclopedia.

Component catalogs sometimes list shift registers as binary ripple **counters**, instead of giving them their own section. In this encyclopedia, a binary counter is considered to have binary-weighted outputs (with values 1, 2, 4, 8…. in decimal notation) and is described in the **counter** entry. A shift register has outputs that are not necessarily binary-weighted.

OTHER RELATED COMPONENTS

- **flip-flop** (see Chapter 11)
- **counter** (see Chapter 13)
- **multiplexer** (see Chapter 16)

What It Does

A *register* is a component (or a small section of computer memory) that stores information. The smallest unit of information is one *bit* (i.e., one *binary digit*) with a value 1 or 0 that can be represented by a high or low logic state. A **shift register** most commonly is designed to store eight bits, although some store four.

Each bit is memorized by the status of a **flip-flop** inside the register. For a detailed description of flip-flops, see Chapter 11. When a pulse from an external *clock* is received by the shift register, all of the bits in storage are moved along one step, from each flip-flop to the next. The high or low status of an input pin at that moment is *clocked in* to the first flip-flop, while the bit in the last flip-flop is overwritten by the bit preceding it. A diagram representing the function of a basic four-bit shift register is shown in Figure 12-1.

Note that the status of the input pin is ignored until the moment when a clock pulse copies it into the first flip-flop. In the figure, when the input pin has a brief high state that ends immediately before clock pulse three, the high state is ignored.

A shift-register chip is shown in Figure 12-2.

Because the functionality of a shift register is now often incorporated in much larger logic chips, it is less widely used as a stand-alone component than it used to be. It is still useful for purposes of serial-parallel or parallel-serial conversion, and for small tasks such as scanning a matrix-encoded keyboard or keypad. It also has educational applications and can be used in conjunction with a microcontroller.

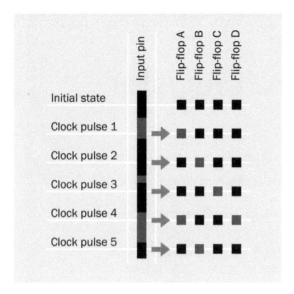

Figure 12-1. *The function of a four-bit shift register in which each flip-flop may be set to a high or a low state, represented here with red or black squares. After a high bit is clocked into the chip, it is moved one space along by each subsequent clock pulse.*

Figure 12-2. *This 8-bit shift-register chip is unusual in that it uses "power logic," in which open-drain outputs enable it to drive relatively high-current devices. It can sink up to 250mA at each of its output pins, at up to 45VDC.*

Schematic Representation

No specific symbol exists for a shift register. It is represented in a schematic by a simple rectangle, often (but not always) with control inputs on the left, data inputs arrayed along the upper edge, and data outputs along the lower edge. An example is shown in Figure 12-3, along with a dia-

gram showing the physical chip and its pinouts. The meaning of the abbreviations identifying the inputs, outputs, and control functions will be described in "How It Works" on page 122.

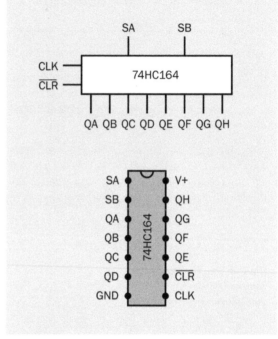

Figure 12-3. *Typical schematic representation of a shift register, compared with the pinouts of the actual component.*

The schematic symbol representing a shift register may appear superficially similar to the physical form of the chip which contains it, but the physical layout of the pins is unlikely to be the same.

How It Works

A shift register generally consists of a chain of D-type flip-flops. See the entry describing **flip-flops** in Chapter 11 for a detailed explanation of this component.

The simplest shift register functions as a *serial-in, serial-out* device, abbreviated with the acronym *SISO*. Because the first bit that enters it will be the first to leave at the opposite end, it can also be described as a *first-in, first-out* data storage device, using the acronym *FIFO*.

The basic connections between flip-flops in a four-bit SISO shift-register are shown in Figure 12-4. The D input in each section refers to the fact that it is a D-type flip-flop. The primary output from each flip-flop is identified with letter Q.

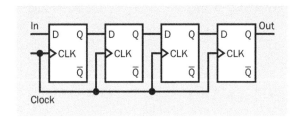

Figure 12-4. *The simplest shift register is a serial-in, serial-out (SISO) device. This example contains four D-type flip-flops.*

Each clock input is labeled CLK. When the output of each flip-flop is coupled to the input of the next flip-flop, and both share the same clock signal, the clock signal will cause the state of the third flip-flop to be sent to the fourth, the output of the second to be copied to the third, the output of the first to the second, and the input state will be copied to the first.

Abbreviations and Acronyms

The shift register will usually have an additional input that forces an immediate "clear" of all the registers, regardless of the clock state at that moment. This input is usually labeled CLR and will have a bar printed above it if it is active-low (which is the usual convention). If there is a pin labeled MR (meaning "Master Reset"), it will have the same function as CLR.

Because its effect is independent of the clock state, the clear signal is described as an *asynchronous* input.

While the abbreviation CLK is frequently used to identify the clock input, SCLK is also used (meaning "serial clock"), and occasionally the abbreviation CP may be found, meaning "clock pulse"

input. If the shift register contains two stages, one to clock data in and the second to clock data out, they may be separately clocked, in which case they will be identified with different abbreviations. These are not standardized, but should be explained in the manufacturer's datasheet. No matter which abbreviation is used for a clock input, it will have a bar printed above it if the input is active-low.

Shift registers are generally *edge triggered*, meaning that the rising or falling edge of a clock pulse triggers the bit-shifting operation. If the component responds to a clock transition from low to high, it is *rising-edge triggered*. If it responds to a transition from high to low, it is *falling-edge triggered*, and this may be indicated in the schematic by a small circle, properly known as a *bubble*, preceding the triangle which indicates that this is an edge-triggered device.

Most shift registers are positive-edge triggered.

Parallel Outputs and Inputs

In many shift registers, data may be read out in parallel (from all flip-flops simultaneously), using pins provided for this purpose. In this mode, the shift register can function as a *serial-parallel converter* (*serial in, parallel out*, represented by the acronym *SIPO*). A simplified schematic of the internal connections is shown in Figure 12-5.

Where parallel outputs are provided, they are often identified as QA, QB, QC, and so on (moving from left to right) but may alternatively be described as Q1, Q2, Q3, Q4, and so on.

In a schematic, the input pin is conventionally shown as being at the left end of the component. Often two inputs are provided, connected internally as inputs to a NAND gate. The inputs are likely to be labeled A and B, but may alternatively be named SA and SB, indicating that they are serial inputs. S1 and S2 are alternative classifications. If parallel inputs exist, they may be identified as PA, PB, PC, and so on.

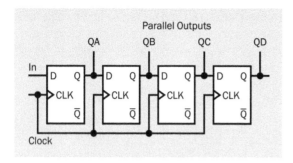

Figure 12-5. *Many shift registers have pins connected to points along the chain of flip-flops. These connections enable data to be read from the shift register in parallel.*

If serial data is supplied asynchronously, as in the illustration in Figure 12-1, it will be ignored until the shift register is triggered by the next clock pulse. The input state at that moment will then be copied into the first flip-flop, while the data that is already being stored in the shift register will be moved along the chain. In datasheets, this is customarily represented by a diagram such as the one in Figure 12-6. This diagram assumes that the shift-register is rising-edge triggered. Note that a brief fluctuation in the input which does not coincide with a clock-trigger event will be ignored.

Variants

Serial In, Serial Out

A basic SISO shift register allows only for serial input (at one end of the chain of flip-flops) and serial output (at the other end of the chain). No pins are available for parallel output of data.

This type of component usually permits 64-bit storage, where parallel output is simply impractical, as too many pins would be required. An example is the 4031B chip. This includes provision for recirculation of bits, so that it will also function as a ring counter (see Chapter 13 for a discussion of this function). As is always the case with logic chips, the part number will be preceded by letter(s) identifying the manufacturer, and a suffix will distinguish variants of the chip.

Another type of SISO shift register is programmable. It will store any number of bits from 1 through 64, determined by a binary number applied in the form of high/low states to five pins reserved for this purpose. An example is the 4557B.

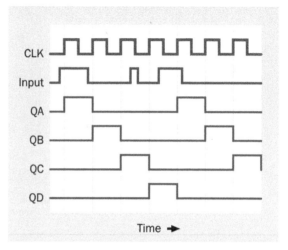

Figure 12-6. *In a rising-edge-triggered shift register, the high or low state of an asynchronous input (purple line) is copied into the first flip-flop of a shift register by each clock pulse (orange line). Brief fluctuations that do not coincide with a rising clock pulse are ignored. Existing data in the register is shifted from one flip-flop to the next.*

Serial In, Parallel Out

The majority of serial-input shift registers allow parallel output from points along the chain in addition to serial output at the end of the chain. These chips almost all are 8-bit registers. Typically two inputs are provided, one of which can be used to receive bits that recirculate from the end of the chain, back to the beginning. Widely used examples are the 4094B and the 74x164, where an acronym identifying the logic family will be substituted for the x.

Parallel In, Serial Out

A minority of shift registers are able to function as *parallel-serial converters* (*parallel in, serial out*, represented by the acronym *PISO*). Typically this type of chip allows *jam-type* parallel data input, meaning that the data is forced into the flip-flops via a separate pin for each of them. Parallel input is enabled by the status of a serial/parallel control pin. When the control pin reverts to its opposite status, each clock pulse will now shift data along the chain of flip-flops, allowing it to be read from the final output one bit at a time. Thus, data can be entered into the chip in parallel and read out of it serially. Examples are the 4014B and 4021B. Both are 8-bit.

Parallel In, Parallel Out

Shift registers that permit parallel output in addition to parallel input are almost all of the *universal* type, described in the next section.

Universal

A *universal* shift register is capable of all four modes of operation: SISO, SIPO, PISO, and PIPO. The four modes of the component are selected by the high or low status of two *mode select* pins. In addition, this component may have the ability to shift the register states either left or right. A *bidirectional* shift register has this same capability, and may also have PISO and PIPO capability, depending on the chip. Examples are the 74x195 and 74x299, where an acronym identifying the logic family will be substituted for the x in the number.

Universal shift registers are almost all 4-bit or 8-bit. They often have relatively complicated features, such as access to internal JK flip-flops, or pins that are multiplexed to provide different functionality depending whether an enable pin is held high or low. Datasheets must be checked carefully to ensure correct use.

Dedicated shift registers of SIPO or PISO type will be easier to use.

Values

As is the case with other logic chips, most flip-flops in the through-hole 74xx series are intended for 5VDC power supply while the older 4000 series may tolerate up to 18VDC. Surface-mount versions may use voltages as low as 2VDC.

See the section on logic gates in Chapter 10 for a discussion of acceptable high and low logic-input states. On the output side, the 4000 series chips are able to source less than 1mA at 5VDC, but the 74HCxx series can usually manage around 20mA.

If a shift register is used for high-speed operation, the following values must be considered (identical notation, and similar values, are found in specifications for flip-flops):

- t_S Setup time: The input state of a shift register must exist for a very brief period before the clock trigger that processes it. This period is known as the *setup time*. In the 4000 series of chips, recommended setup may be as long as 120ns. The value will be much lower in 74xx chips.

- t_H Hold time: The minimum time in nanoseconds for an input to persist after the active edge of a clock pulse that has processed it. In many shift registers, no hold time is necessary, as the chip has already been activated by the rising edge of the clock pulse.

- t_{CO} Clock-to-output: The elapsed time after a clock trigger, before the output changes. This is a function of the internal workings of the chip, and may be broken down into low-to-high and high-to-low output transitions, as follows.

- T_{PLH} Propagation to Low-to-High: The elapsed time after an active clock trigger, before a low-to-high swing occurs at an output. This may not be identical to T_{PHL}.

- T_{PHL} Propagation to High-to-Low: The elapsed time after an active clock trigger, before a high-to-low swing occurs at an output. This may not be identical to T_{PLH}.

- f_{MAX} Maximum clock frequency for reliable operation. In the older design of 4000 series chips, 3MHz may be recommended with a power supply of 5VDC. Higher frequencies are possible with a higher voltage power supply. Frequencies as high as 20MHz are possible in the 5VDC 74HC00 series.

- $t_{W(H)}$ The minimum high clock pulse width in nanoseconds. In the older design of 4000 series chips, 180ns may be recommended with a power supply of 5VDC. Shorter pulses are possible with higher voltage power supply. Pulses as short as 20ns are possible in the 5VDC 74HC00 series.

- $t_{W(L)}$ The minimum low clock pulse width in nanoseconds. This is likely to be the same as $t_{W(H)}$.

Power Considerations

Shift registers conform with the usual power-supply requirements for logic families. These are described in detail in the **logic gate** entry in Chapter 10. Likewise, the ability of a shift register to source or sink current is usually determined by its logic family. In a few cases, however, shift registers have an additional *open-drain* output stage for each register, capable of sinking currents as high as 250mA. The Texas Instruments TPIC6596 shown in Figure 12-2 is an example. When an open-drain output is connected with a logic device whose input cannot be allowed to float indeterminately, a *pullup resistor* must be added.

Three-State Output

A chip may have a *three-state output* (also known as *tri-state output*, a term which was trademarked but is now used generically). This means it will be capable of changing its outputs so that instead of sourcing or sinking current in a logical high or low state, they can have a high impedance. The chip then becomes "invisible" to others which may be sharing the same output bus. The high-impedance state is usually applied to all outputs from the shift register simultaneously when enabled by a separate pin, often identified as *OE*, meaning *output-enable*. Examples of three-state shift registers are the 74x595 or the 4094B chip.

The high-impedance state can be thought of as being almost equivalent to switching the out-

puts of the shift register out of the circuit. Consequently, if other components sharing the bus are also in high-impedance output mode, the state of the bus will "float," with uncertain results. To avoid this, a pullup resistor of 10K to 100K can be used on each bus-line.

Where the internal components of a shift register are shown in a datasheet, a three-state output is usually represented with a buffer symbol or inverter symbol that has an additional control input located on its upper edge, as shown in Figure 12-7. This should not be confused with the similar placement of a positive power supply input to an amplifier or op-amp. (Schematics showing the interior elements of a logic chip almost never include power-supply connections.)

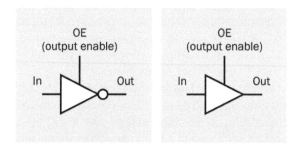

Figure 12-7. *A shift register may be capable of a three-state output, where high impedance is an option additional to the usual high or low logic state. An output enable pin allows this. It is typically shown as an additional input to an inverter (left) or buffer (right) inside the shift-register chip.*

How to Use It

The SISO application of a shift register can be used simply to delay the transmission of data by storing it and moving it from one flip-flop to the next before it is read out of the end of the chain.

The SIPO application of a shift register (serial in, parallel out) may be useful where a microcontroller has insufficient outputs to control multiple devices. Serial data can be sent from a single microcontroller output to the input of a shift register. The chip can then drive a separate device from each of its parallel output pins. If there are eight devices, the microcontroller can send a sequence of eight bits, each of which will control the on/off status of one device when the bits are read out of the shift register in parallel. If more devices are used, an additional shift-register can be daisy-chained to the output from the first.

Clock signals can be supplied from the microcontroller, along with a signal to the clear input of the shift register, if desired. Alternatively, the old bit states in the shift register can simply be "clocked out" and replaced with a new set of serial data. During the process of "bit banging," the parallel outputs of the shift register can remain connected directly with the output devices if the clock speed from the microcontroller is fast enough, as devices such as relays will not respond to extremely brief pulses.

For generic shift registers that do not have open-drain outputs, a buffer will be needed to provide sufficient current for any device using more current than an LED.

If a shift register is configured for PISO mode (parallel in, serial out) it can be placed on the input side of a microcontroller, polling a variety of devices and feeding their states into the microcontroller serially.

Dual Inputs

Where a shift register has two serial inputs (as is often the case), they are almost always linked as inputs to an internal NAND gate. This allows the output from the end of the chain of flip-flops to be connected back to the beginning of the chain, if the shift register is to function as a ring counter. However, if this function is not used and a single input is required, the two inputs to the shift register can be tied together. In this configuration, the input becomes active-low. The two possible configurations are shown in Figure 12-8. It is important never to leave an input *floating*, or unconnected.

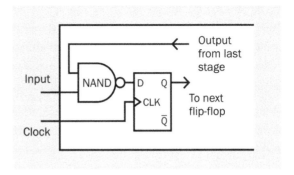

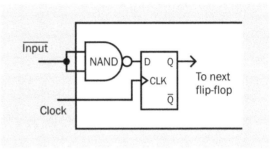

Figure 12-8. *Two possible configurations where a shift register allows two inputs linked with an internal NAND gate.*

Preloading the Shift Register

Where a shift register will be used to output a single recirculating bit (or in similar applications), the component must be preloaded with a high state in its first register. This may require a one-shot (monostable) timer which is activated only when the circuit is powered up.

Polling a Keyboard

Two shift registers can be used to scan the data lines in a matrix-encoded keyboard or keypad. This is often known as *polling* the keyboard. Provided the scan rate is sufficiently fast, the user experiences a seemingly immediate response to a key-press.

While the full schematic is too large and complex to be included here, many examples can be found online.

Arithmetical Operations

Shift registers were traditionally used to perform arithmetical operations on binary numbers. If the number is represented by eight bits (i.e., one byte) with the most significant digit on the left, shifting the bits one space to the right will have the effect of dividing the byte value by 2. If the bits are shifted one space to the left, the byte value will be multiplied by 2 (provided an additional register is available to store the most significant bit after it has been shifted). This concept is illustrated in Figure 12-9.

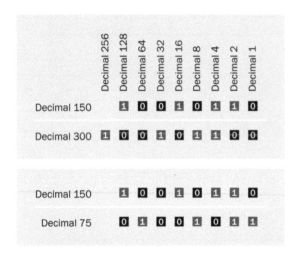

Figure 12-9. *In the upper section of this diagram, a binary number represented by eight bits in a shift register is multiplied by 2 by shifting all the bits one space to the left. In the lower section, the same binary number is divided by 2 by shifting all the bits one space to the right. The binary values are shown in decimal notation to the left.*

In the upper section of the figure, the binary number 10010110 (chosen arbitrarily) is represented in the eight flip-flops of a shift register. The decimal place value of each digit is indicated. Adding up the place values of the 1s in the number, the total is 128 + 16 + 4 + 2 = 150. Below the white line, the digits have been shifted one place to the left, with the leftmost digit carried over to an additional location, while a zero is inserted in the rightmost location. Assuming that the additional place at the leftmost location has a place value of 256, the total is now 256 + 32 + 8 + 4 = 300.

In the lower section of the figure, the same binary number has been shifted one space to the right,

with a 0 introduced in the leftmost location. The decimal value is now 64 + 8 + 2 + 1 = 75.

While this application for shift registers was common during development of digital computing in the 1960s and 1970s, the shift register as a separate component became less common subsequently, as its functionality was acquired by modern CPU chips.

Buffering

A shift register may also be used as a *buffer* between two circuits where the clock speeds are different. Digits are clocked in by the first circuit, then clocked out by the second. Some shift registers allow two clock inputs and can be used for this purpose.

What Can Go Wrong

Problems that tend to affect digital chips generally are listed in the entry on logic gates (see "What Can Go Wrong" on page 105).

Confusing Classification

Because of the functional similarity to a binary ripple **counter**, a shift register is sometimes listed by component suppliers as a counter. In fact, a binary counter will almost always have outputs that have place values 1, 2, 4, 8…. and upward, while the outputs from a shift register will not have place values.

When searching for a shift register, it can be found by specifying a "counting sequence" of serial to parallel, serial to serial, parallel to serial, or parallel to parallel. If the "counting sequence" is simply up or down, the component is a counter, not a shift register.

Inadequate Setup Time

Each flip-flop in a shift register must have a stable input state before the next triggering event shifts the data. If this setup time is reduced below the minimum specified in the datasheet, results will be unpredictable.

Unconnected Input

Because many shift registers have a choice of two inputs to the same chain of internal flip-flops, it is easy to leave one of them unconnected by accident. A floating input will be vulnerable to any stray electromagnetic fields, and is almost certain to create random effects.

Output Enable Issues

The output enable pin on a shift register that has three-state logic outputs is usually active-low. Consequently, if the pin is left unconnected, the logic outputs may go into high-impedance mode, or will fluctuate unpredictably. Where three-state outputs are not required, a safe course of action will be to avoid using three-state chips.

Floating Output Bus

If a pullup resistor is omitted from a bus that is shared by three-state chips, the results will be unpredictable. Even if the circuit design seems to guarantee that at least one chip will have a high or low output on the bus, a pullup resistor should still be included.

counter | 13

The term **counter** is used here to mean a digital-logic chip. A counter could be built from discrete transistors, but this approach is obsolete. Counters may also be devised from parts such as multiple relays, or a solenoid advancing a ratchet wheel, but such electromechanical devices are not included in this encyclopedia.

In this encyclopedia, a counter by definition has binary-weighted outputs with values 1, 2, 4, 8…. in decimal notation. The exception to this rule is a *ring counter,* which does not have binary-weighted outputs but is included here because its name identifies it as a counter. A **shift register** may be used as a ring counter, but is more versatile and has many other functions; hence it has a separate entry.

Gray code counters, in which successive outputs differ by only one binary digit, are not described in this encyclopedia.

OTHER RELATED COMPONENTS

- **flip-flop** (see Chapter 11)
- **shift register** (see Chapter 12)

What It Does

A counter can be used to count events, or can measure time in convenient intervals if it is connected with a component such as a *quartz crystal* that operates at a precise and reliable frequency. The counter receives input pulses (usually referred to as a *clock input*) and counts a predetermined number of them before restarting from the beginning. It will repeat in this fashion so long as power is connected, and the clock pulses continue, and a reset signal is not supplied.

Almost all counters create some form of output during the count. Most commonly, the output is a pattern of high and low states expressing the number of clock pulses in binary code. Where a counter counts to a very high number before recycling, some intermediate binary digits may be omitted.

While standalone counter chips are not as widely used now as in the early days of computing, they still find application in industrial processes, small devices, and education, and can be used to control incremental devices such as stepper motors. They can be used in conjunction with microcontrollers.

Schematic Representation

No specific logic symbol exists for a counter. It is most often shown in a schematic as a rectangle with clock input(s) and clear input(s) on the left and outputs on the right. An example appears in Figure 13-1, above a representation of the physical chip and its pinouts. The meaning of the abbreviations identifying the inputs, outputs, and control functions will be found in "How It Works" on page 132. Because the two MR inputs for this particular counter are ANDed inside the chip, the AND symbol is included with the counter symbol.

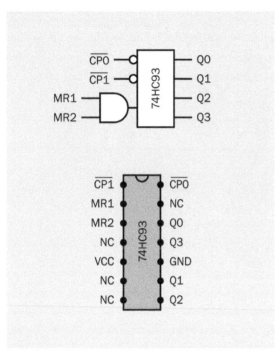

Figure 13-1. *Typical schematic representation of a counter, compared with the pinouts of the actual component.*

A counter chip is shown in Figure 13-2.

Figure 13-2. *The 74HC163 shown in this photograph is a 4-bit synchronous counter capable of being preloaded with a starting value, and able to do a synchronous reset.*

How It Works

A counter is built from a chain of **flip-flops**, with each one triggering the next. JK, T-type, or D-type flip-flops may be used. For a thorough description of a flip-flop, see Chapter 11. In Figure 13-3, a D-type flip-flop is shown, triggered by each rising clock pulse.

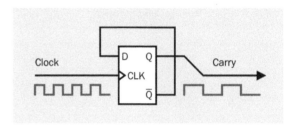

Figure 13-3. *When the complementary (NOT-Q) output from a D-type flip-flop is connected back to its input, the Q output frequency is half of the clock input frequency. See text for additional details.*

Initially the Q output of the flip-flop is low, so its NOT-Q output (identified by Q with a bar above it) is high. This feeds back to the D input, but has no effect until the rising edge of the next clock pulse copies the high D input to the Q output. The Q output is now latched high while the NOT-Q output is latched low and feeds back to the D input. The triggering event has passed, so the low D input does not have any immediate effect. The rising edge of the next clock pulse copies the low state of the D input to the output, and changes the NOT-Q output to high, causing the cycle to repeat. As a result, the output frequency of the flip-flop is one-half of the input frequency from the clock. If the output is carried to the next flip-flop to become its clock input, once again the frequency will be divided by two.

Modulus and Modulo

The *modulus* of a counter is the number it will count up to, before it repeats. This is sometimes written incorrectly as *modulo*.

In fact, *modulo* is the name of an arithmetical operation, often abbreviated as *MOD* (usually capitalized, even though it is not an acronym). This operation consists of a division sum in which the *remainder* from the division is the result of the operation. Thus 100 modulo 5 gives a result of 0, because there is no remainder when 100 is divided by 5. But 100 modulo 7 gives a result of 2, because 2 is the remainder of the division operation.

To exacerbate the confusion, *MOD* is also used as an adjective referring to the modulus of a counter. Thus a MOD-4 counter has a modulus of 4, and a MOD-16 counter has a modulus of 16. As a general rule, when a counter is being described, *modulo* and *MOD* will mean the same thing as *modulus*. This may be confusing to people such as computer programmers who are already familiar with the correct usage of MOD as an arithmetical operator.

In a counter, to obtain a modulus that is not a power of two, logic gates inside the chip can intercept a particular value (such as 1010 binary, which is 10 decimal) and use this as a signal to restart the count at zero. External connections to the chip can achieve the same purpose.

Pin Identifiers

Abbreviations and acronyms are used in datasheets to identify pin functions. These identifiers have not been standardized, and many variants exist.

CLK is the abbreviation most commonly used for the clock input, sometimes alternatively shown as CK or CP. If it is active-low, or if its falling edge will be active, a bar will be printed above it. Where a printed font does not permit an underscore to be placed in this way, CLK' (the abbreviation followed by an apostrophe) may be used instead. Often two or more stages of a counter can be separately clocked, and the input pins will be identified with abbreviations such as CLK1 and CLK2, or 1CLK and 2CLK, or CKA and CKB, or CP1 and CP2, or similar.

Where a clock input is edge-triggered, this is indicated by a small triangle. The triangle can be seen in Figure 13-3.

CLR designates a pin which will clear the count and reset it back to zero. This signal is often active-low, indicated by a bar printed above the abbreviation.

In a schematic, a small circle, properly known as a *bubble*, may be placed at any input which is active-low. On a clock input, the circle indicates that it is falling-edge triggered. See Figure 11-16 for various implementations of symbols with flip-flop schematics.

The CLR operation may be *synchronous* (the pin state will not be recognized until the next clock pulse) or *asynchronous* (in which case the pin state overrides the clock and resets the counter immediately). MR stands for "Master Reset," and has the same function as CLR.

Where two or more counters (or multiple stages within one counter) can be reset separately, more than one clear input will be shown, and may be identified with abbreviations such as CLR1 and CLR2, or MR1 and MR2.

Output pins will almost always be identified as Q0, Q1, Q2 … or QA, QB, QC … up to the maximum necessary to express the modulus. If two or more counters are included in one chip, outputs may be prefixed with an appropriate number. Thus 2Q3 would be the third output in the second counter. Multiple counters in one chip are identified with numbers starting from 1.

Where internal flip-flops are shown, they will have identifiers such as FF1 or FF2. Each flip-flop will have its own clear function, identified as C or CD (the latter meaning "clear data"), and may have inputs labeled D1, D2, D3… in a D-type flip-flop or J and K in a JK flip-flop. See Chapter 11 for an explanation of flip-flop inputs and outputs.

The input to a counter is always imagined to begin from the left, and consequently the leftmost flip-flop shown in an internal schematic will express the least significant bit of the current value of the count, even though a binary number is written with the least significant bit in the rightmost place.

If a counter is capable of receiving parallel data as an input (explained below) it will have a pin labeled PE for parallel enable. It may also have a CE or CET pin, for count enable.

As is generally the case in logic chips, VCC or V+ are typically used to identify the positive power supply pin, while GND or V- will identify the negative-ground pin. NC means that a pin has no internal connection at all, and consequently requires no external connection.

Variants

All counter chips use binary code internally, and the number of *bits* (*binary digits*) in the counter's modulus will be the same as the number of internal flip-flops. A 4-bit counter (the usual minimum) will have a modulus of 2^4 which is 16. A 21-bit counter (the maximum typically available) will have a modulus of 2^{21} which is 2,097,152. For higher moduli, counters can be chained together, each sending a carry signal to the next. This is known as a *cascade*.

Multiple counters, with different moduli, may be chained in a single chip. For example, in a digital clock that displays hours and minutes using the 60Hz frequency of an American domestic power supply as its timebase, the initial counting stage will have a modulus of 60, to count individual seconds. The next counting stages will have moduli of 60, 10, and 6, so that they can count from 00 to 59 minutes. Additional stages in the chip will tally hours.

A counter with a *parallel input* can be preloaded with an initial value (in binary code) from which it may count up or down. A *parallel-enable* pin may put the counter into a mode where the number can be *jam loaded*, meaning that it is jammed into the counter regardless of the clock state. Other types of counter are loaded synchronously.

Ripple versus Synchronous

In a *ripple counter* each internal flip-flop triggers the clock input of the next, so that their states change in a rapid but incremental sequence from the input to the output. This is also known as an *asynchronous* counter. Because the final state will not be valid until the clock pulse has rippled all the way through the counter (and through additional counters if they are cascaded together), a ripple counter will tend to suffer from a *propagation delay* of up to a microsecond. Ripple counters may also create output spikes or momentary transient count values that are invalid. Therefore, they are more suitable for applications such as driving a numeric display than for interfacing at high speed with other logic chips.

In a *synchronous counter*, all the flip-flops are clocked simultaneously. A synchronous counter is better suited to operation at high speed.

Of the counter chips available today, about half are synchronous and half are asynchronous.

Ring, Binary, and BCD

A counter that activates output pins one at a time sequentially is said to have a *decoded output*. It is often referred to as a *ring counter*. It has the same number of output pins as its modulus. An example is the 4017B chip.

A *binary counter* is more common and has an *encoded output*, meaning that it will express the running total of the count in binary code through *weighted outputs* that typically have (decimal) values of 1, 2, 4, 8, and so on. A modulus-8 counter (often referred to as an *octal counter*) will require three outputs which represent the binary numbers 000, 001, 010, 011, 100, 101, 110, and 111 (decimal 0 through 7) before going back to 000 and repeating.

A *modulus-16* counter, also known as a *hexadecimal counter* or a *divide-by-16 counter*, will have a binary output represented by four output pins, counting from 0000 through 1111 (decimal 0 through 15). Four-digit binary counters are very common, and their outputs are compatible with other components such as a **decoder**, which converts a binary-number input into a ring-counter-style output.

A *decade counter* is a modulus-10 binary counter. It is described as having a *binary-coded decimal* output (often expressed with the acronym *BCD*), using four weighted output pins which represent the numbers 0000, 0001, 0010, 0011, 0100, 0101, 0110, 0111, 1000, and 1001 (decimal 0 through 9) before repeating. Because this counter skips binary outputs from 1010 through 1111 (decimal 10 through 15), it is said to have a *shortened modulus*.

Figure 13-4 shows a schematic diagram of JK flip-flops in a decade ripple counter. The J and K inputs are all tied to the positive power supply, as this causes the clock input to toggle the output high and low. Note that because the primary input is always shown at the left end of the component, the least significant output bit (Q0) is in the leftmost position.

To intercept binary 1010 (decimal 10), an internal NAND gate is used. Its output goes low when its two inputs, from Q1 and Q3, go high. The output from the NAND immediately activates the CLR function on all the flip-flops, so that as soon as the decade counter reaches 1010 (decimal 10), it resets itself to 0.

In this particular chip, the preload for each flip-flop is tied to the positive power supply, so that it is always inactive. In some counters, the preload feature of each flip-flop is accessible via pins outside the chip. This creates the potential hazard of preloading the counter with one of the numbers that it normally skips (for instance, 11 decimal in a decade counter). This is referred to as an *invalid number* or *disallowed state*. (This use of the term "state" refers only to the binary number stored in the counter's flip-flops. It has nothing to do with the high-state or low-state voltages used to represent binary 0 or 1.)

The counter's datasheet should include a *state diagram* showing how the counter will deal with this situation. It may reset itself to a valid value after a maximum of two steps, but this can still cause confusion, depending on the application.

The state diagram for a 74HC192 counter is shown in Figure 13-5.

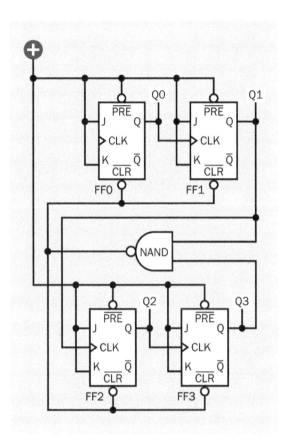

Figure 13-4. *The internal logic of a synchronous decade counter that uses JK flip-flops.*

Clock Sources

The clock input may be provided by a **timer** chip or by an *RC network*, which has the advantage of being able to run at a relatively low speed for purposes where this is desirable. It may alternatively be provided by a *quartz crystal* oscillating at a much higher frequency such as 1MHz. Successive counters may be necessary to reduce this value, depending on the application.

In some counters, the clock is built into the chip. More commonly, a resistor and capacitor may be used externally to establish a clock rate in conjunction with logic gates inside the chip. The datasheet for this type of component will include a formula for calculating the clock frequency from the resistor and capacitor values. The 4060B chip is an example.

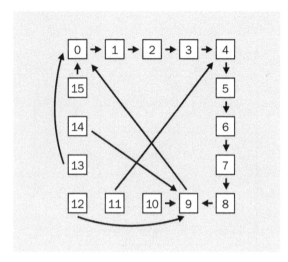

Figure 13-5. *A state diagram shows the transitions that a counter will make from each number to the next (in decimal notation), including the transitions which enable it to exit from disallowed states. This example is for a 74HC192 chip.*

Rising Edge and Falling Edge

A counter may be designed to be triggered either by the rising edge or the falling edge of the clock input, or by its high or low logic state. Generally speaking, ripple counters use the falling edge, so that the final output from one counter can become the clock input of the next. In other words, when the most significant digit of the first counter changes from a high to low logic state, this transition toggles the least significant bit of the second counter.

Synchronous counters generally use the rising edge of the clock input. If multiple synchronous counters are cascaded, they must all share the same clock signal, and will all change their flip-flop states simultaneously.

Multiple Stages

It is common for a counter chip to contain two or more *stages* with differing moduli. To take a common example, a divide-by-2 stage and a divide-

by-5 stage that are both present in a single chip can be used to create a decade counter by connecting external pins. The extra stages provide a choice of moduli if they are used individually.

Single and Dual

Counter chips may contain two counters of the same modulus. This is known as a *dual* counter. Dual 4-bit counter chips are common. Each counter can be used separately, or they can be cascaded, in which case the total modulus will be found by multiplying the individual moduli together.

High-State, Low-State, and Three-State

Almost all counters use positive logic where a 1 is represented by a high state and 0 by a low state. Some counters allow an additional output state which has a high impedance and is equivalent to an open circuit. This feature is useful when two or more chips share the same output bus. It is discussed in the entry for **shift registers** in "Three-State Output" on page 126.

Descending Output

Most components only create an ascending count. The output can be converted to a descending count by passing each binary state through an inverter, but this will only work properly if the modulus is equal to the number of states. In a BCD counter, its inverted outputs will count from decimal 15 to decimal 6, not from decimal 9 to decimal 0.

A few counters are available which are designed to create a correct descending count. Other counters are available which allow the user to set the mode to ascending or descending. Examples are the 74x190 or 74x192 (where an acronym for the chip family will be substituted for the letter x).

A descending output is useful in combination with a parallel input, where a user may set an initial value from which the counter will descend to 0. With suitable logic, this can enable a user-specified delay period.

Programmable Counters

A programmable counter can usually allow a modulus ranging from 2 to more than 10,000. The counter counts down by dividing an initial number repeatedly with a value that is preset with binary inputs. An example is the 4059B chip.

Examples

Many counter specifications date back to the 4000 family of logic chips. Versions of them subsequently became available in the 74xx series, often with the old 4000 part number preceded by 74x (where x is replaced by a designation of the logic family). For example, a version of the 4518B dual BCD chip can be obtained as the 74HC4518. As is the case with all logic chips, this part number will be preceded by letter(s) designating the particular manufacturer, with a suffix added to differentiate minor variants of the chip. The 74xx series has the advantage of higher speed and a greater ability to source or sink current at its output pins.

Most of the original CMOS chips, such as the 4518B, are still available, even in surface-mount versions. These offer the possible advantage of being able to use a higher power supply voltage.

Many counters offer multiple options such as different modulus values that can be selected by external pin connections. Some chips are tolerant of slow clock frequencies; others are not. Most are edge-triggered, but a few are level triggered. Some, such as the 4518B mentioned above, allow a choice of a rising-edge clock input and falling-edge clock input on different pins. For a specific application, it is really necessary to read a variety of datasheets to select the chip that is most suitable.

Values

As is the case with other logic chips, most counters in the through-hole 74xx series are intended for 5VDC power supply while the older 4000 series may tolerate up to 18VDC. Surface-mount 74xx versions may use voltages as low as 2VDC.

See the section on logic gates in Chapter 10 for a discussion of acceptable high and low logic-input states. On the output side, the 4000 series chips are able to source or sink less than 1mA at 5VDC, but the 74HCxx series can usually manage around 20mA.

A few counters are capable of delivering more power through additional output stages that can drive LEDs. The 4026B decade counter is still being manufactured, capable of powering modest 7-segment displays. The 4033B has the additional option of blanking any leading zeros in a multidigit display. Other chips that were designed for direct connection to LED numerals have become obsolete as the need for this application has diminished. The 74C925, 74C926, 74C927, and 74C928 are examples. They may still be found from surplus outlets, but should not be specified in new circuit designs.

What Can Go Wrong

The entry that deals with problems affecting shift registers (see "What Can Go Wrong" on page 128), describes issues which also affect counters. The entry that deals with logic chips (see "What Can Go Wrong" on page 105), describes problems affecting all types of logic chips. In addition, the potential problems listed below are specific to counters.

Lock-Out

This is the condition which occurs if a counter with a shortened modulus is loaded with a binary state that is out of its range. Consult the datasheet and study its state diagram to determine the most likely outcome if this problem occurs.

Asynchronous Artifacts

Because the flip-flops in an asynchronous (ripple) counter do not change simultaneously, they create very brief false outputs while the ripple process is taking place. In a 4-bit counter, the binary number 0111 (decimal 7) should be followed by 1000 (decimal 8). However, the rightmost digit (i.e., the least significant bit) will change to a 0 initially, creating 0110 as a momentary binary output (decimal 6). The carry operation will then change the next digit to a 0, creating 0100 (decimal 4). The carry operation continues, changing the next digit to a 0, creating 0000. Finally the operation completes by creating 1000 as the correct output.

These intermediate states on the output pins are often referred to as *glitches*. Because they are extremely brief, they will be indetectable when a counter is used to drive a display. They can cause significant issues, however, if the outputs of the counter are connected with other logic chips.

Another type of asynchronous problem will occur if the clock speed is sufficiently high that a new pulse is received at the first flip-flop before the ripple of changing states has passed all the way through to the final flip-flop. This will result in a different brief invalid value on the output pins.

Noise

Old TTL-type counters, such as the 74LSxx series, are especially noise-sensitive. Adding a 0.1µF or 0.047µF bypass capacitor as close to the power supply pin as possible is recommended. Breadboarding counters of this type may result in errors if a high-frequency clock is used, because conductors such as patch-cords are liable to pick up noise. Modern 74HCxx counters are preferable.

encoder

14.

In this encyclopedia, an **encoder** is a digital chip that converts a decimal-valued input into a binary-coded output.

The term "encoder" may alternatively refer to a **rotational encoder** (also known as a *rotary encoder*) which has a separate entry in Volume 1 of this encyclopedia. The term may also describe a *code hopping encoder*, which is an encryption device used in keyless entry systems for automobiles.

OTHER RELATED COMPONENTS

- **decoder** (see Chapter 15)
- **multiplexer** (see Chapter 16)

What It Does

An encoder is a *logic chip* that receives an input consisting of an active logical state on one of at least four input pins, which have decimal values from 0 upward in increments of 1. The encoder converts the active pin number into a binary value represented by logic states on at least two output pins. This behavior is opposite to that of a **decoder**.

Encoders are identified in terms of their inputs and outputs. For example:

- 4-to-2 encoder (four input pins, two output pins)
- 8-to-3 encoder (eight input pins, three output pins)
- 16-to-4 encoder (sixteen input pins, four output pins)

In the early days of computing, encoders processed interrupts. This application is now rare, and relatively few encoder chips are still being manufactured. However, they are still useful in small devices—for example, if a large number of inputs must be handled by a microcontroller that has insufficient pins to receive data from each individually.

Schematic Symbol

Like other logic-based components, the encoder does not have a specific schematic symbol and can be represented by a plain rectangle as in Figure 14-1, with inputs on the left and outputs on the right. The bars printed above some of the abbreviations indicate that an input or output is active-low. In this chip, the 74LS148, all inputs and outputs are active-low.

Generally speaking, inputs labeled D0, D1, D2… are used for data input, although they may simply be numbered, with no identifying letter. The encoded outputs are typically identified as Q0, Q1, Q2… or A0, A1, A2… with Q0 or A0 designating the least significant bit in the binary number.

Pins labeled E and GS are explained in the following section.

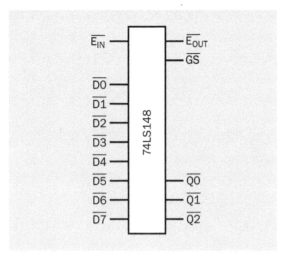

Figure 14-1. *While no specific schematic symbol exists for an encoder chip, this style is commonly used. Shown here is a 16–to–4 encoder with active-low inputs and outputs.*

Similar Devices

The similarities and differences between encoder, decoder, multiplexer, and demultiplexer can cause confusion.

- In an **encoder**, an active logic state is applied to one of four or more input pins, while the rest remain in an inactive logic state. The input pin number is converted to a binary code which is expressed as a pattern of logic states on two or more output pins.

- In a **decoder**, a binary number is applied as a pattern of logic states on two or more input pins. This value determines which one of four or more output pins will have an active logic state, while the rest remain in an inactive logic state.

- A **multiplexer** can connect a choice of multiple inputs to a single output, for data transfer. The logic state of an enable pin, or a binary number applied as a pattern of logic states to multiple control pins, chooses which input should be connected with the output pin. The alternative term *data selector* evokes the function of this device more clearly.

- An analog multiplexer may allow its inputs and outputs to be swapped, in which case it becomes a **demultiplexer**. It can connect a single input to one of multiple outputs, for data transfer. The logic state of an enable pin, or a binary number applied as a pattern of logic states to multiple control pins, chooses which output should be used. The alternative term *data distributor* evokes the function of this device more clearly.

How It Works

An encoder contains **logic gates**. The internal logic of an 8-to-3 encoder is shown in Figure 14-2, where the darker blue rectangle represents the chip. The switches in this figure are external and are included only to clarify the concept. An open switch is imagined to provide an inactive logic input, while a single closed switch provides an active logic input. (Multiple active inputs can be handled by a *priority encoder*, described below).

Each input switch has a numeric status from 1 to 7. The switch with value 0 does not make an internal connection, because the output from the OR gates is 000 by default.

The logic state of each OR output represents a binary number, weighted with decimal values 1, 2, and 4, as shown at the bottom of the figure. Thus, if switch 5 is pressed, by tracing the connections it is clear that the outputs of OR gates 4 and 1 become active, while the output from gate 2 remains inactive. The values of the active outputs thus sum to 5 decimal.

Figure 14-3 shows the outputs for all possible input states of a 4-to-2 encoder. Figure 14-4 shows the outputs for all possible input states of an 8-to-3 encoder. These diagrams assume that a high logic state is an active logic state, on input or output. This is usually the case.

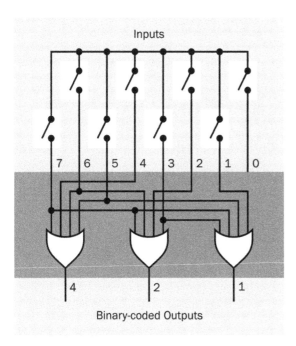

Figure 14-2. *A simplified simulation of the internal logic of an 8-to-3 encoder. The dark blue rectangle indicates the space inside the chip. The external switches are included only to clarify the concept. An encoder chip would have an Enable line to create an active output.*

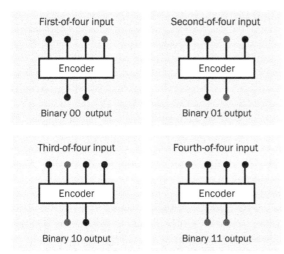

Figure 14-3. *The four possible inputs of a 4-to-2 encoder (top of each panel) and the encoded outputs (below).*

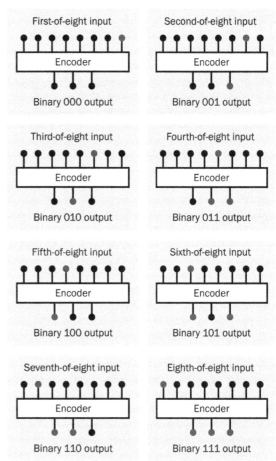

Figure 14-4. *The eight possible inputs of an 8-to-3 encoder (at the top of each panel), and the encoded outputs (below). Note that one input of an encoder must always be logic-high. All logic-low inputs are not a valid state.*

Unlike ripple counters, where propagation delays can reduce the overall response time of the component, decoders respond within two or three nanoseconds.

Variants

A *simple encoder* assumes that only one input pin can be logically active at a time. A *priority encoder* assigns priority to the highest-value input pin if more than one happens to receive an active input. It ignores any lower-value inputs. An example is the 74LS148, which is an 8-to-3 chip.

A few encoders feature *three-state* outputs (also known as *tri-state*), in which a high-impedance or "floating" output state is available in addition to the usual high and low logic states. The high-impedance state allows multiple chips to share an output bus, as those that are in high-impedance mode appear to be disconnected. This is useful if two or more encoders are cascaded to handle a larger number of inputs.

Values

As is the case with other logic chips, most encoders in the through-hole 74xx series are intended for 5VDC power supply while the older 4000 series may tolerate up to 18VDC. Surface-mount versions may use voltages as low as 2VDC.

See the section on logic gates in Chapter 10 for a discussion of acceptable high and low input states. On the output side, the 4000 series chips are able to source less than 1mA at 5VDC, but the 74HCxx series can manage around 20mA.

How to Use It

Suppose that a microcontroller should respond to an eight-position rotary switch. Because the switch cannot be turned to more than one position at a time, all of its eight contacts can be connected with the inputs on an encoder, which will deliver a 3-bit binary number to three inputs of the microcontroller. Code inside the microcontroller then interprets the pin states.

This is shown in Figure 14-5. Pulldown resistors would be needed on the input pins of the encoder, to prevent them from floating when they are not connected by the rotary switch. They have been omitted from this diagram for simplicity. Debouncing the switch would be handled by the microcontroller.

Other forms of input may be used instead of a rotary switch. For example, the outputs from eight comparators or eight phototransistors could be passed through an encoder.

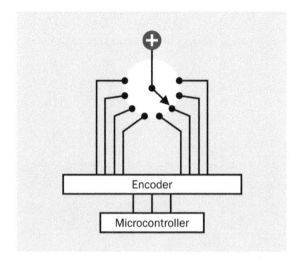

Figure 14-5. *Output from an eight-input rotary switch could be connected through an 8-to-3 encoder to provide input to a microcontroller using a reduced number of pins. Pulldown resistors have been omitted for simplicity.*

Cascaded Encoders

Encoders are often provided with features to facilitate handling additional inputs via multiple chips. Typically, a second Enable pin is provided, as an output that connects with the Enable input of the preceding chip. This preserves the priority function, so that an input on the second chip prevents any additional input to the first chip from affecting the output. In a datasheet, the enable pins may be labeled E_{IN} and E_{OUT}, or E_I and E_O.

In addition, a GS pin will be included, meaning "Group Select." It is logically active only when the encoder is enabled and at least one input is active. The GS pin of the most-significant encoder provides an additional binary digit.

The outputs from two encoders can be linked via OR gates, as shown in Figure 14-6, where the lower chip's GS output provides the most significant bit of a four-bit binary number.

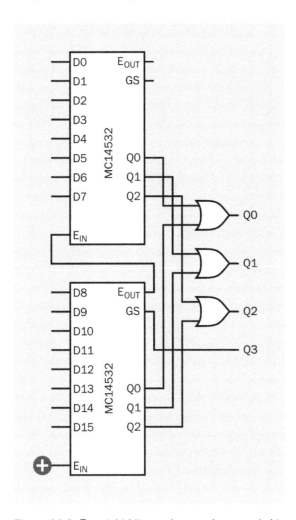

Figure 14-6. *Two eight-bit encoders can be cascaded to handle 16 separate inputs. In this example, the encoders use active-high logic.*

What Can Go Wrong

Problems that are common to all digital chips are summarized in the section on logic gates in "What Can Go Wrong" on page 105.

See "What Can Go Wrong" on page 149 in the entry describing **decoders** for a list of more specific problems that also afflict encoders.

decoder | 15

In this Encyclopedia, a **decoder** is a digital chip that receives a binary-coded input and converts it to a decimal output by applying a logic state to one of a sequence of pins, each of which is assigned an integer value from 0 upward.

The term "decoder" also refers to components and devices that have other functions, such as decoding audio or video formats. These functions are not included here.

OTHER RELATED COMPONENTS

- **encoder** (see Chapter 14)
- **multiplexer** (see Chapter 16)

What it Does

A decoder receives a binary-coded number on two or more input pins. It *decodes* that number and expresses it by activating one of at least four output pins.

The behavior of a decoder with a two-bit binary input is shown in four sequential snapshots in Figure 15-1, where the least significant bit of the input is on the right in each diagram, and the output moves from right to left.

Figure 15-2 shows a similar sequence in a decoder where various values of a three-bit input are decoded to create an eight-pin output.

One sample state of a four-bit decoder is shown in Figure 15-3.

All of these figures assume that a high state represents an active input or output. In a few chips, a low state is used to represent an active output.

Decoders with 2, 3, or 4 input pins are common. To handle a binary input greater than 1111 (decimal 15), decoders can be chained together, as described below.

Manufacturers' datasheets often describe decoders in terms of their inputs and outputs. Typical examples would include:

- 2-to-4 decoder (two input pins, four output pins)
- 3-to-8 decoder (three input pins, eight output pins)
- 4-to-10 decoder (for converting *binary-coded decimal* to decimal output)
- 4-to-16 decoder (also known as a *hex decoder*).

Input Devices

The input pins of a decoder can be driven by a counter that has a binary-coded output. A decoder can also be driven by a microcontroller, which may have an insufficient number of output pins to control a variety of devices. Two, three, or four of the outputs can be used to represent a binary number which is passed through the decoder to activate the devices one at a time, perhaps with transistors or Darlington arrays introduced to handle the load. This is suggested in Figure 15-4.

A **shift register** can be used for a similar purpose, but often has only one pin for input. This pin must be supplied sequentially with a serial pattern of bits that will match the desired high/low states of the output pins. The relative advantage of this system is that a shift register can generate any pattern of output states. A one-of-many decoder can activate only one output at a time.

LED Driver

A special case is a *seven-segment decoder* designed to drive a seven-segment **LED display** numeral. A binary-coded decimal number on four input pins is converted to a pattern of outputs appropriate for lighting the segments of the display that will form a number from decimal 0 through 9.

Schematic Symbol

Like other logic-based components, the decoder does not have a specific schematic symbol and is represented by a plain rectangle as in Figure 15-5, with inputs on the left and outputs on the right. The bars printed above the E and LE abbreviations (which stand for Enable and Latch Enable, respectively) indicate that they are active-low. In this chip, the 74HC4514, all outputs are active-high, but in a related 4-to-16 decoder, the 74HC4515, all outputs are active-low. In both of these chips, the Enable pin is held low to activate the outputs. The Latch Enable pin freezes the current state of the outputs (i.e., it latches them) when it is held low.

Generally speaking, pins labeled A0, A1, A2... in a datasheet are often the binary inputs (although A, B, C... may be used), with A0 designating the least significant bit. Outputs are usually labeled Y, and are activated in sequence from Y0 when the binary input starts counting upward.

Similar Devices

The similarities and differences between encoder, decoder, multiplexer, and demultiplexer can cause confusion.

- In a **decoder**, a binary number is applied as a pattern of logic states on two or more input pins. This value determines which one of four or more output pins will have an active logic state, while the rest remain in an inactive logic state.

- A **multiplexer** can connect a choice of multiple inputs to a single output, for data transfer. The logic state of an enable pin, or a binary number applied as a pattern of logic states to multiple control pins, chooses which input should be connected with the output pin. The alternative term *data selector* evokes the function of this device more clearly.

- An analog multiplexer may allow its inputs and outputs to be reversed, in which case it becomes a **demultiplexer**. It can connect a single input to one of multiple outputs, for data transfer. The logic state of an enable pin, or a binary number applied as a pattern of logic states to multiple control pins, chooses which output should be used. The alternative term *data distributor* evokes the function of this device more clearly.

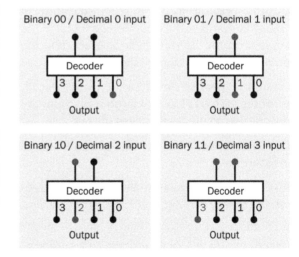

Figure 15-1. *A decoder with two input pins can interpret their binary-number representation to create an active logic state on one of four output pins.*

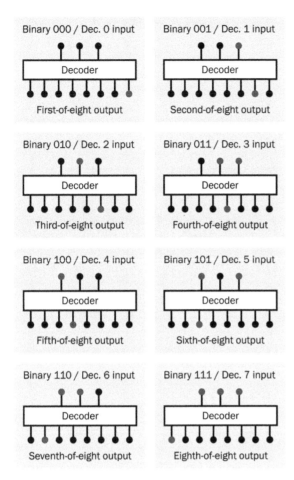

Figure 15-2. *A decoder with three input pins can interpret their binary-number representation to create a high logic state on one of eight output pins.*

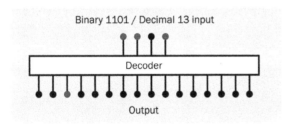

Figure 15-3. *A decoder with four input pins can interpret their binary-number representation to create a high logic state on one of 16 output pins. Only one of the 16 possible states is shown here.*

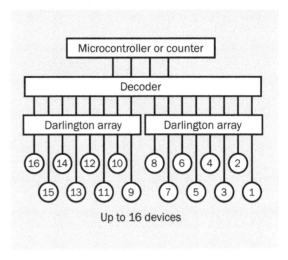

Figure 15-4. *Four outputs from a binary counter or micro-controller can be used by a decoder to activate one of up to 16 output devices.*

A photograph of a 74HC4514 decoder chip appears in Figure 15-6.

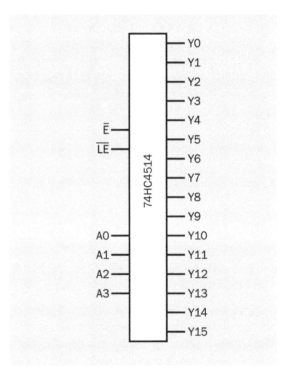

Figure 15-5. *While no specific schematic symbol exists for a decoder chip, this style is commonly used. Shown here is a 4-to-16 decoder.*

Figure 15-6. *The 24-pin 74HC4514 decoder chip process-es a 4-bit input and represents it by making one of its 16 output pins active-high.*

How It Works

A decoder contains logic gates, each of which is wired to respond to a unique binary pattern of inputs. (In the case of a seven-segment decoder, the internal logic is more complicated.) Figure 15-7 shows the logic of a 2-to-4 decoder. The darker blue area contains the components inside the chip. The external switches are included only to clarify the function of the decoder. An open switch is imagined to provide a low logic input, while each closed switch provides a high logic input.

Unlike ripple counters, where propagation delays can reduce the overall response time of the component, decoders function within two or three nanoseconds.

Variants

Decoder variants have not proliferated with time, and relatively few are available. Most are 3-to-8, 4-to-16, and binary-coded-decimal types.

The 7447 and 74LS47 are seven-segment decoders that have an open-collector output capable of driving a 7-segment display directly. The 7448 is similar but also contains built-in resistors and a capability to blank out leading zeros in a display. However, some suppliers now list the 74LS48 as obsolete. It may be still available from

old stock, but should not be specified in new circuits.

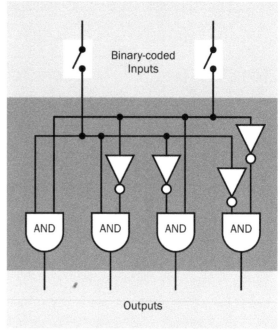

Figure 15-7. *A simplified simulation of the logic in a decoder. An actual chip would have an Enable line to activate the output. The dark blue rectangle indicates the space inside the chip.*

Although 74LS47 is still being manufactured, and is available in surface-mount as well as through-hole format, a version is not available in the widely used HC family of 74xx chips. Care must be taken to satisfy the input voltage requirements of the 74LS47 when driving it with 74HCxx chips.

Values

As is the case with other logic chips, most decoders in the through-hole 74xx series are intended for 5VDC power supply while the older 4000 series may tolerate up to 18VDC. Surface-mount versions may use voltages as low as 2VDC.

See the section on logic gates in Chapter 10 for a discussion of acceptable high and low input states. On the output side, the 4000 series chips are able to source or sink less than 1mA at 5VDC, but the 74HCxx series can manage around 20mA.

How to Use It

The original applications for decoders in computer circuits have become uncommon, but the chips can still be useful in small appliances and gadgets where multiple outputs are controlled by a **counter** or microcontroller.

Although 16 is usually the maximum number of outputs, some chips are designed to allow expansion. The 74x138 (where a chip family identifier such as LS or HC can be substituted for the letter x) is a 3-to-8 decoder with two logic-low Enable pins and one logic-high Enable. If a value-8 binary line is applied to the low-enable of one chip and the high-enable of another, the first chip will be disabled when the line goes high to indicate that the binary number 1000 has been reached, and the second chip can continue upward from there by sharing the same three less-significant-bit inputs. As many as four chips can be chained in this way.

What Can Go Wrong

Problems that are common to all digital chips are summarized in the section on logic gates in "What Can Go Wrong" on page 105.

Glitches

Although a decoder typically functions faster than a ripple counter, it suffers the same tendency to introduce brief *glitches* in its output. These are momentary invalid states which occur while processes inside the chip that are slightly slower are catching up with other processes that reach completion slightly faster. A brief *settling time* is necessary to ensure that the output is stable and valid. This will be irrelevant when powering a device such as an **LED indicator**, which will not display such brief transients. The problem may be more important if the output from the decoder is used as an input to other logic chips.

If the input to a decoder is derived from a ripple counter, the input may also contain glitches, which can cause erroneous outputs from the decoder. It is better to use a synchronous counter on the input side of a decoder.

Unhelpful Classification

Online parts suppliers tend to list decoders under the same category heading as encoders, multiplexers, and demultiplexers, making it difficult to find what you want. Under this broad subject heading (which will include thousands of chips), if you search by selecting the number of inputs relative to the number of outputs that you have in mind, this will narrow the search considerably.

Active-Low and Active-High

Chips with identical appearance and similar part numbers may have outputs that are either active-low or active-high. Some may offer a latch-enable pin, while others have enable pins that must be pulled low or forced high to produce an output. Accidental chip substitution is a common cause of confusion.

multiplexer 16

May be abbreviated as a *mux* (this term is sometimes printed all in caps), and may be referred to alternatively as a *data selector*. Some sources maintain that a multiplexer has no more than two channels, whereas a data selector has more, but there is no consensus on this, and datasheets continue to use the term "multiplexer" predominantly.

Analog multiplexers are usually bidirectional, and thus will function equally well as *de-multiplexers*. Consequently, this encyclopedia does not contain a separate entry for demultiplexers.

OTHER RELATED COMPONENTS

- **encoder** (see Chapter 14)
- **decoder** (see Chapter 15)

What It Does

A multiplexer can select one of two or more input pins, and connect it internally with an output pin. Although it is an entirely solid-state device, it behaves as if it contains a **rotary switch** in series with a SPST switch, as shown in Figure 16-1. A binary code applied to one or more Select pins chooses the input, and an Enable pin establishes the connection with the output. The Select and Enable functions are processed via an internal section referred to as a *decoder*, not to be confused with a **decoder** chip, which has its own entry in this encyclopedia.

All multiplexers are digitally controlled devices, but may be described as either *digital* or *analog* depending how they process the input signal. A digital multiplexer creates an output that is adjusted to logic-high or logic-low within the limits of its logic family. An analog multiplexer does not impose any processing on the voltage, and passes along any fluctuations. Thus, it can be used with alternating current.

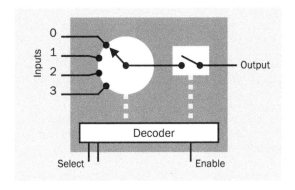

Figure 16-1. *A multiplexer functions as if it contains a rotary switch. The switch position is determined by a binary number applied to external Select pins. The internal connection is completed by applying a signal to an Enable pin.*

Because an analog multiplexer merely switches a flow of current, it can be *bidirectional*; in other words, it can function as a *demultiplexer*, in which case the input is applied to the pole of the (imaginary) internal switch and outputs are taken from the terminals.

151

Differential Multiplexer

A *differential* multiplexer contains multiple switches that are differentiated from one another (i.e., they are electrically isolated, although they are controlled by the same set of select pins). A differential multiplexer is conceptually similar to a rotary switch with two or more decks controlled by a single shaft. See Figure 16-2.

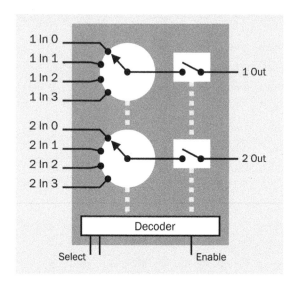

Figure 16-2. *A differential multiplexer contains two or more electronic switches that are differentiated from one another, similarly to the decks on a rotary switch. Although the channels into each switch are typically numbered from 0 upward, the switches are numbered from 1 upward.*

A bidirectional dual 4-channel differential analog multiplexer is shown in Figure 16-3.

Modern multiplexers are often found switching high-frequency data streams in audio, telecommunications, or video applications.

Similar Devices

The similarities and differences between multiplexer, demultiplexer, encoder, and decoder can cause confusion:

- A **multiplexer** can connect a choice of multiple inputs to a single output, for data transfer. The logic state of an enable pin, or a binary number applied as a pattern of logic

states to multiple control pins, chooses which input should be connected with the output pin. The alternative term *data selector* evokes the function of this device more clearly.

- An *analog* multiplexer may allow its inputs and outputs to be reversed, allowing it to become a **demultiplexer**, connecting a single input to one of multiple outputs, for data transfer. The logic state of an enable pin, or a binary number applied as a pattern of logic states to multiple control pins, chooses which output should be used. The alternative term *data distributor* evokes the function of this device more clearly.

Figure 16-3. *This CMOS chip contains two four-channel differential analog multiplexers.*

- In an **encoder**, an active logic state is applied to one of four or more input pins, while the rest remain in an inactive logic state. The input pin number is converted to a binary code which is expressed as a pattern of logic states on two or more output pins.

- In a **decoder**, a binary number is applied as a pattern of logic states on two or more input pins. This value determines which one of four or more output pins will have an active logic state, while the rest remain in an inactive logic state. A digital multiplexer does not allow reversal of its inputs and outputs, but a decoder functions as if it were a digital demultiplexer.

How It Works

The multiple inputs to a multiplexer are referred to as *channels*. Almost always, the number of channels is 1, 2, 4, 8, or 16. A 1-channel component is only capable of "on" or "off" modes and functions similarly to a SPST switch.

If there are more than two channels, a binary number will determine which channel is connected internally. The number of channels is usually the maximum that can be identified by the number of select pins, so that 2 pins will control 4 channels, 3 pins will control 8 channels, and 4 pins (the usual maximum) will control 16 channels.

In multiplexers with three or more channels, an enable pin is usually still present to activate or deactivate all the channels simultaneously. The enable feature may be described alternatively as a *strobe*, or may have an inverse function as an *inhibit* pin.

Although a rotary switch is helpful in conceptualizing the function of a multiplexer, a more common representation (sometimes in datasheets) is an array of SPST switches, each of which can be opened or closed by the decoder circuit. A typical example, depicting a dual differential multiplexer, is shown in Figure 16-4. Note that the internal decoder can only close one switch in each channel at a time.

The switch analogy is appropriate in that when an output from a multiplexer is not connected internally (i.e., its switch is "open") it is effectively an open circuit. However, some multiplexers contain *pullup resistors* to give each output a defined state. This can be an important factor in determining whether the multiplexer is suitable for a particular application.

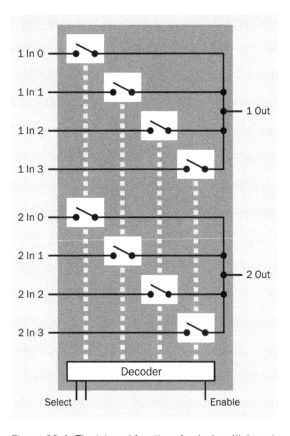

Figure 16-4. *The internal function of a dual multiplexer is commonly represented as a network of SPST switches, each of which is controlled by decoder logic.*

A digital multiplexer actually contains a network of logic gates, shown in simplified form in Figure 16-5.

A demultiplexer has internal logic shown in simplifier form in Figure 16-6.

Schematic Symbol

In a schematic, a multiplexer and demultiplexer may be represented by a trapezoid with its longer vertical side oriented toward the larger number of connections. This is shown in Figure 16-7. However, this symbol is falling into disuse.

More often, as is the case with most logic components, a multiplexer or demultiplexer is represented by a rectangle with inputs on the left and outputs on the right, as shown in Figure 16-8. The distinction between inputs and outputs is problematic, however, in an analog multiplexer which will allow data flow to be reversed.

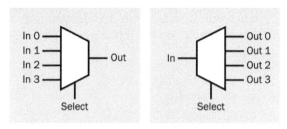

Figure 16-7. *The traditional symbol for a multiplexer (left) and demultiplexer (right). The trapezoid is oriented with its longer vertical side facing the larger number of connections. This symbol is falling into disuse.*

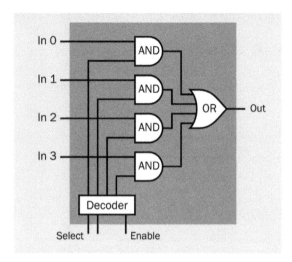

Figure 16-5. *A simplified representation of the logic gates in a digital multiplexer.*

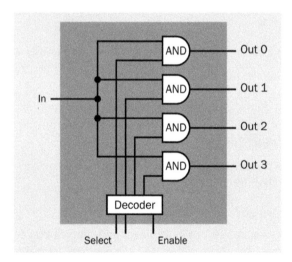

Figure 16-6. *A simplifier representation of the logic gates in a digital demultiplexer.*

Pin Identifiers

The lack of standardization in the identification of pin functions is perhaps more extreme in the case of multiplexers than for other types of logic chips.

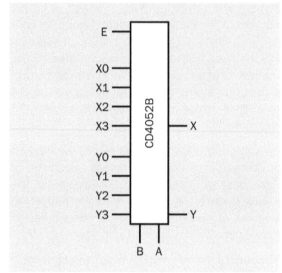

Figure 16-8. *A simple rectangle is most often used as a schematic symbol for a multiplexer, but the abbreviations assigned to pin functions are not standardized. See text for details.*

An output enable pin will be shown as E or EN, or occasionally OE. It may alternatively be described as an inhibit pin, labeled INH, or sometimes will be called a strobe. The function is the same in each case: one of its logic states will enable the internal switches, while its other logic state will prevent any internal switches from closing.

Switch inputs may be labeled S0, S1, S2… or X0, X1, X2… or may simply be numbered, almost always counting up from 0. Where two or more sets of switches coexist in one package, each set of inputs may be distinguished from the others by preceding each identifier with a numeral or letter to designate the switch, as in 1S0, 1S1, 1S2... or 1X0, 1X1, 1X2... (Switches are generally numbered from 1 upward, even though their inputs are numbered from 0 upward.) Outputs may be identified using the same coding scheme as inputs, bearing in mind that the inputs and outputs of an analog multiplexer usually are interchangable. Some manufacturers, however, prefer to identify each multiplexer output by preceding it with letter Y. Alternatively, Z1, Z2, Z3... may identify the outputs from switches 1, 2, 3... Fortunately, datasheets usually include some kind of key to this grab-bag of abbreviations.

Control pins are often identified as A, B, C... with letter A representing the least significant bit in the binary number that is applied to the pins.

Voltages can be confusing in multiplexers. Components intended for use with digital inputs are straightforward enough, as the supply voltage will be identified as V_{CC} and is typically 5VDC for through-hole packages (often lower for surface-mount), while negative ground is assumed to be 0VDC. However, where a multiplexer may be used with AC inputs in which the voltage varies above and below 0V, supply voltages above and below 0VDC are also possible—such as +7.5VDC and −7.5VDC, to take a random example. Three power-supply pins may be provided for this purpose. The positive supply will usually be identified as V_{DD} (the D refers to the Drain in the internal MOSFETs). A V_{EE} pin may be at 0VDC or at a negative value equal and opposite to V_{DD}. The E in this abbreviation is derived from Emitter voltage, even though the component may not contain a bipolar transistor with an emitter. Customarily, a V_{SS} pin (the S being derived from the Source in the internal MOSFETs) will be at 0VDC, and other voltages will be measured above and below this baseline. This ground pin may alternatively be labeled GND.

As is customary in logic chips, low-active control pins will have a bar printed above their identifi-

ers, or an apostrophe will be placed after an identifier if the font does not permit printing the bar. Alternatively, low-active pins may be represented by showing a small circle, properly referred to as a *bubble*, at the input or output point of the symbol for the multiplexer. Note that analog inputs and outputs are neither high-active nor low-active; they merely pass voltages through.

Variants

Most multiplexers are "break before make" devices, where one input is disconnected before the next input is connected. However, some exceptions exist, and datasheets should be checked for this. It can be a significant issue, because make-before-break switching will briefly connect external devices with each other, through the chip.

Many multiplexers can tolerate control voltages above the usual high value in a logic circuit—as high as 15VDC in some cases. The voltage that is switched by the multiplexer may be the same as the control voltage, or may be higher.

Some analog multiplexers have *overvoltage protection* that allows them to withstand input voltages that are twice or three times the recommended maximum.

Datasheets may mention "internal address decoding," meaning that the binary number input, specifying a channel to be switched, is decoded inside the chip. In fact, virtually all multiplexers now have on-chip address decoding, and this feature should be assumed to exist, regardless of whether it is mentioned.

Values

The voltage to be switched will usually be referred to as the *input voltage*, V_{IN}.

An analog multiplexer should not be subjected to current exceeding the value that it is designed to switch. This is known as the maximum *channel current*. A typical value would be 10mA, although many modern surface-mount components are designed for currents in the microamp range.

The *on-resistance* is the resistance imposed by the analog multiplexer on the signal flowing through it. While modern, specialized analog multiplexers may have an on-resistance as low as 5Ω, these are relatively unusual. An on-resistance of 100Ω to 200Ω is more common. This value will vary within a component depending on the power supply voltage and the voltage being switched. It will increase slightly as V_{IN} deviates above (or below) 0V, will increase substantially for lower values of supply voltage, and will increase significantly with temperature.

The curves in Figure 16-9 show on-resistance of an analog multiplexer varying with input voltage, with three different power supplies: plus-and-minus 2.5VDC (described in the graph as a "spread" of 5VDC), plus-and-minus 5VDC (a "spread" of 10VDC), and plus-and-minus 7.5VDC (a "spread" of 15VDC). These curves were derived from a datasheet for the MC14067B analog multiplexer; curves for other chips will be different, although the basic principles remain the same.

Switching time is an important consideration in high-speed applications. The "on" and "off" times specified in a datasheet (often as t_{ON} and t_{OFF}) are a function of the propagation delay from the control input to the toggling of the switch, and are generally measured from the halfway point of the rising or falling edge of the control input, to the 90% point of the output signal level.

Leakage current is the small amount of current (often measured in picoamperes) that the solid-state switch will pass when it is in its "off" state. This should be insignificant except when very high-impedance loads are used.

Separate switches inside a multiplexer may have characteristics that differ slightly from one another. Differences in on-resistance between adjacent switches can be important when switching parallel analog signals. A datasheet should mention the extent to which switches have matched characteristics, and may define the maximum deviation from one another using the abbreviation R_{ON} even though this same term

may be used, confusingly, to denote the on-resistance of each individual switch.

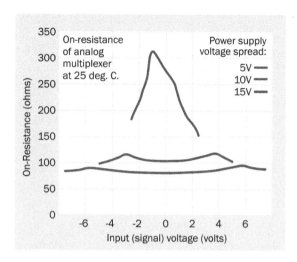

Figure 16-9. *Variations in on-resistance in an analog multiplexer. Each voltage "spread" is the difference between positive supply voltage and an equal-and-opposite negative ground voltage. Thus a "spread" of 10VDC means plus and minus voltages of 5VDC. (Curves derived from On Semiconductor datasheet for MC14067B analog multiplexer.)*

How to Use It

A multiplexer may be used as a simple switch to choose one of multiple inputs, such as a choice of input jacks on a stereo system. A dual differential multiplexer is useful in this application, as it can use a single select signal to switch two signal paths simultaneously.

A multiplexer can also be used as a digital volume control by switching an audio signal among a variety of resistances, similar to a **digital potentiometer**. In this application, the possible presence of pullup resistors inside the multiplexer must be considered.

Where a microcontroller must monitor a large number of inputs (for example, a range of temperature sensors or motion sensors), a multiplexer can reduce the number of input pins required. Its data-select pins will be cycled through all the possible binary states by the microcontroller, to select each data input in turn, while its single-wire output will carry the analog data to a separate pin on the microcontroller which performs an analog-digital conversion.

Conversely, a demultiplexer (i.e., an analog multiplexer such as the 4067B chip which can be used in demultiplexer mode) can be used by a microcontroller to switch multiple components on and off. Four outputs from the microcontroller can connect with the control pins of a 16-channel demultiplexer, counting from binary 0000 through binary 1111 to select output pins 0 through 15. After selecting each pin, the microcontroller can send a high or low pulse through it. The process then repeats. (A **decoder** can be used in the same way.)

Other Application Notes

Multiplexers may be *cascaded* to increase the inputs-to-outputs ratio.

Modern multiplexers are found on computer boards where they choose among video output ports, or as PCI express channel switches.

A multiplexer may be used as a parallel-to-serial converter, as it samples multiple channels and converts them into a serial data stream.

In telecommunications, a multiplexer can sample voice signals from multiple separate inputs and combine them into a digital stream that can be transmitted at a faster bit rate over a single channel. However, this application goes far beyond the simple uses for multiplexers described here.

What Can Go Wrong

Problems that are common to all digital chips are summarized in the section on logic gates (see "What Can Go Wrong" on page 105).

Pullup Resistors

While they are often necessary to prevent connections from floating, pullup resistors built into a multiplexer may have unexpected consequences if the user is unaware of them.

Break Before Make

For most applications, it is desirable for each internal solid-state switch to break one connection before making a new one. This avoids the possibility of separate external components being briefly connected with each other through the multiplexer. Datasheets should be checked to verify that a multiplexer functions in break-before-make mode. If it doesn't, the enable pin can be used momentarily to disable all connections before a new connection is established.

Signal Distortion

Where a multiplexer is passing analog signals, signal distortion can result if the on-resistance of multiple internal switches varies significantly at different voltages. A datasheet for an analog multiplexer should usually include a graph showing on-resistance over the full signal range. The flatter the graph is, the less distortion the component will create. This is often described in datasheets as R_{ON} Flatness.

Limits of CMOS Switching

Although most multiplexers are built around CMOS transistors, their switching speed may be insufficient for video signals, and their on-resistance may vary enough to introduce distortion. Multiplexers are available with complementary bipolar switching for very high-speed applications. They impose some penalties in cost and power consumption.

Transients

Switch capacitance inside a multiplexer can cause transients in the output when the switch changes state. An allowance for *settling time* may be necessary. This will be additional to the switching speed claimed by the datasheet.

LCD 17

The full term *liquid-crystal display* is seldom used. Its acronym, **LCD**, is much more common. Sometimes the redundant combination **LCD display** is found. All three terms refer to the same device. In this encyclopedia, the first two words in *liquid-crystal display* are hyphenated because they are an adjectival phrase. Other sources often omit the hyphen.

The acronym **LED** (for **light-emitting diode**) is easily confused with **LCD**. While both devices display information, their mode of action is completely different.

OTHER RELATED COMPONENTS

- **LED display** (see Chapter 24)

What It Does

An LCD presents information on a small display panel or screen by using one or more segments that change their appearance in response to an AC voltage. The display may contain alphanumeric characters and/or symbols, icons, dots, or pixels in a *bitmap*.

Because of its very low power consumption, a basic monochrome LCD is often used to display numerals in battery-powered devices such as digital watches and calculators. A small liquid-crystal display of this type is shown in Figure 17-1.

Color-enabled, backlit LCDs are now frequently used in almost all forms of video displays, including those in cellular telephones, computer monitors, game-playing devices, TV screens, and aircraft cockpit displays.

How It Works

Light consists of electromagnetic waves that possess an electric field and a magnetic field. The fields are perpendicular to each other and to the direction in which the light is traveling, but the field polarities are randomly mixed in most visi-

ble radiation. This type of light is referred to as *incoherent*.

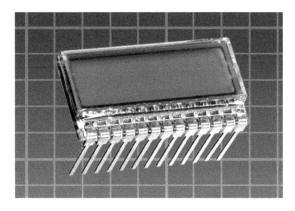

Figure 17-1. *A small, basic monochrome LCD.*

Figure 17-2 shows a simplified view of an LCD that uses a backlight. Incoherent light emerges from the backlight panel (A) and enters a vertical polarizing filter (B) that limits the electric field vector. The polarized light then enters a liquid crystal (C) which is a liquid composed of molecules organized in a regular helical structure that rotates the polarity by 90 degrees when no voltage is applied to it. The light now passes through

159

a horizontal polarizing filter (D) and is visible to the user.

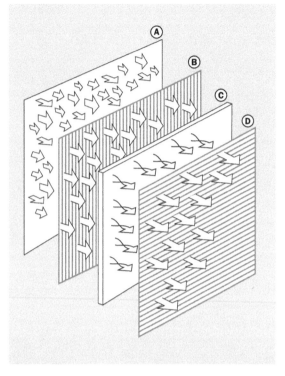

voltage must be used. An AC frequency of 50Hz to 100Hz is common.

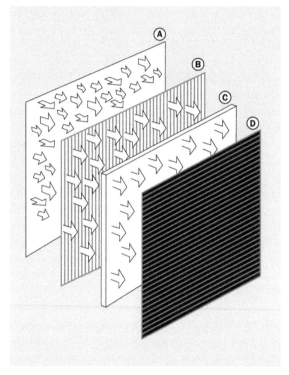

Figure 17-2. *The combination of two polarizers and a liquid crystal appears transparent when voltage is not applied. See text for details.*

Figure 17-3. *The LCD appears dark when voltage is applied. See text for details.*

- A liquid crystal itself does not emit light. It can only modify light that passes through it.

Figure 17-3 shows what happens when voltage is applied to the liquid crystal via transparent electrodes (not included in the figure). The molecules reorganize themselves in response to the electric potential and allow light to pass without changing its polarity. Consequently, the vertically polarized light is now blocked by the front, horizontally polarized filter, and the display becomes dark.

A liquid crystal contains ionic compounds that will be attracted to the electrodes if a DC voltage is applied for a significant period of time. This can degrade the display permanently. Therefore, AC

Variants

A *transmissive LCD* requires a backlight to be visible, and is the type illustrated in Figure 17-2. In its simplest form, it is a monochrome device, but is often enhanced to display full color by adding red, green, and blue filters. Alternatively, instead of a white backlight, an array of pixel-sized red, green, and blue **LEDs** may be used, in which case filters are unnecessary.

Backlit color LCDs have displaced *cathode-ray tubes*, which used to be the default system in almost all video monitors and TVs. LCDs are not only cheaper but can be fabricated in larger sizes. They do not suffer from *burn in*, where a persistent unchanging image creates a permanent scar in the phosphors on the inside of a tube. However, large LCDs may suffer from dead pixels or stuck pixels as manufacturing defects. Different manufacturers and vendors have varying policies regarding the maximum acceptable number of pixel defects.

In a *reflective LCD*, the structure is basically the same as that shown in Figure 17-2 except that a reflective surface is substituted for the backlight. Ambient light enters from the front of the display, and is either blocked by the liquid crystal in combination with the polarizing filters, or is allowed to reach the reflective surface at the rear, from which it reflects back through the liquid crystal to the eye of the user. This type of display is very easily readable in a bright environment, but will be difficult to see in dim conditions and will be invisible in darkness. Therefore, it may be augmented with a user-activated light source mounted at the side of the display.

A *transreflective LCD* contains a translucent rear polarizer that will reflect some ambient light, and is also transparent to enable a backlight. While this type of LCD is not as bright as a reflective LCD and has less contrast, it is more versatile and can be more energy efficient, as the backlight can be switched off automatically when ambient light is bright enough to make the display visible.

Active and Passive Types

An *active matrix* LCD adds a matrix of thin-film transistors to the basic liquid-crystal array, to store the state of each segment or pixel actively while the energizing AC voltage transitions from positive to negative. This enables a brighter, sharper display as *crosstalk* between adjacent pixels is reduced. Because *thin-film transistors* are used, this is often described as a *TFT* display; but the term is interchangable with *active matrix*.

A *passive matrix* LCD is cheaper to fabricate but responds sluggishly in large displays and is not so well suited to fine gradations in intensity. This type of component is used primarily in simple monochrome displays lacking intermediate shades of gray.

Crystal Types

Twisted Nematic (TN) are the cheapest, simplest type of LCD, allowing only a small viewing angle and average contrast. The appearance is limited to black on gray. The response rate is relatively slow.

Super Twisted Nematic (STN) displays were developed in the 1980s for passive LCDs, enabling better detail, wider view angle, and a faster response. The natural appearance is dark violet or black on green, or dark blue on silver-gray.

Film-compensated Super Twisted Nematic (FSTN) uses an extra coating of film that enables a pure black on white display.

Double Super Twisted Nematic provides further enhancement of contrast and response times, and automatic contrast compensation in response to ambient temperature. The appearance is black on white. This display requires backlighting.

Color Super Twisted Nematic (CSTN) is an STN display with filters added for full color reproduction.

Seven-Segment Displays

The earliest monochrome LCDs in devices such as watches and calculators used seven segments to display each numeral from 0 through 9. This type of LCD is still used in low-cost applications. A separate control line, or electrode, connects to each segment, while a *backplane* is shared by all the segments, connecting with a *common* pin to complete the circuit.

Figure 17-4 shows a typical seven-segment display. The lowercase letters *a* through *g* that identify each segment are universally used in datasheets. The decimal point, customarily referred to as "dp," may be omitted from some displays. The array of segments is slanted forward to enable more acceptable representation of the diagonal stroke in numeral 7.

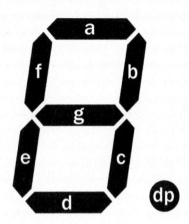

Figure 17-4. *Basic numeric display format for LCD numeric displays (the same layout is used with LEDs). To identify each segment, lowercase letters are universally used.*

Seven-segment displays are not elegant but are functional and are reasonably easy to read, as shown in Figure 17-5. Letters A, B, C, D, E, and F (displayed as A, b, c, d, E, F because of the restrictions imposed by the small number of segments) may be added to enable display of hexadecimal values.

In appliances such as microwave ovens, very basic text messages can be displayed to the user within the limitations of 7-segment displays, as suggested in Figure 17-6.

The advantage of this system is low cost, as 7-segment displays are cheap to fabricate, entail the fewest connections, and require minimal decoding to create each alphanumeric character. However, numbers 0, 1, and 5 cannot be distinguished from letters O, I, and S, while letters containing diagonal strokes, such as K, M, N, W, X, and Z, cannot be displayed at all.

Figure 17-5. *Numerals and the first six letters of the alphabet created with a 7-segment display.*

Additional Segments

Alphanumeric LCDs were developed using 14 or 16 segments to enable better representation of letters of the alphabet. Sometimes these displays were slanted forward, like the 7-segment displays, perhaps because the style had become familiar, even though the addition of diagonal segments made it unnecessary. In other cases, the 14 or 16 segments were arrayed in a rectangle. See Figure 17-7.

The same words represented in Figure 17-6 are shown in Figure 17-8, using 16-segment LCDs. Clearly, the advantage gained by enabling diagonal strokes entailed the disadvantage of larger gaps in the letters, making made them ugly and difficult to read.

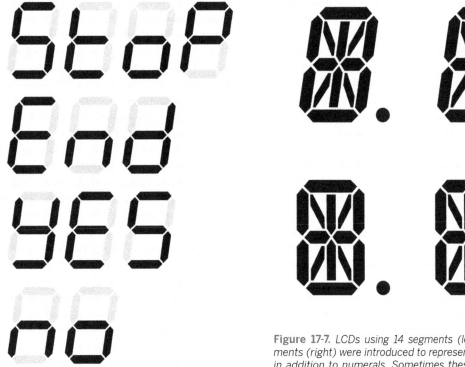

Figure 17-6. *Basic text messages can be generated with 7-segment displays, although they cannot contain alphabetical letters that use diagonal strokes.*

A full character set using 16-segment LCDs is shown in Figure 17-9. This conforms partially with the ASCII coding system, in which each character has an identifying numeric code ranging from 20 hexadecimal for a letter-space to 7A hexadecimal for letter z (although this character set does not attempt to represent lowercase letters differently from uppercase). The ASCII acronym stands for *American Standard Code for Information Interchange*.

Because backlit LCDs had become common by the time 16-segment displays were introduced, the characters were often displayed in light-on-dark or "negative" format, as suggested in this figure. LEDs, of course, have always used the light-on-dark format, as an LED is a light-emitting component.

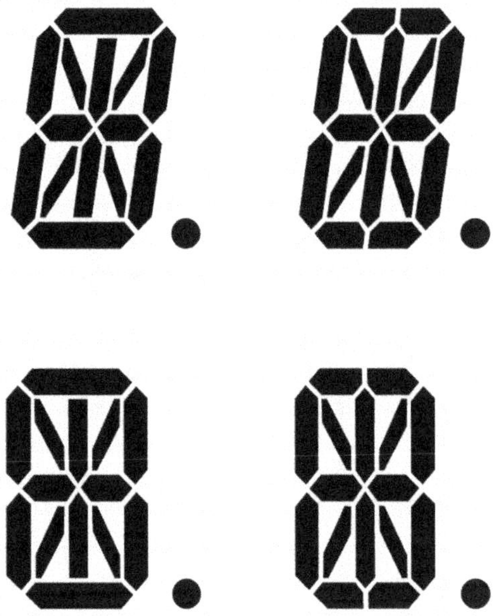

Figure 17-7. *LCDs using 14 segments (left) and 16 segments (right) were introduced to represent a full alphabet in addition to numerals. Sometimes these displays were slanted forward, like the previous 7-segment type, even though this was no longer necessary to represent the number 7.*

Dot-Matrix Displays

The 16-segment displays were never widely popular, and the declining cost of microprocessors, LCD fabrication, and ROM storage made it economic to produce displays using the more easily legible 5x7 dot-matrix alphabet that had been common among early microcomputers. Figure 17-10 shows a dot-matrix character set that is typical of many LCDs.

Because the original ASCII codes were not standardized below 20 hexadecimal or above 7A hexadecimal, manufacturers have represented a variety of foreign-language characters, Greek letters, Japanese characters, accented letters, or symbols using codes 00 through 1F and 7B through FF. The lower codes are often left blank, allowing user installation of custom symbols. Codes 00 through 0F are often reserved for control functions, such as a command to start a new line of text. There is no standardization in this area, and the user must examine a datasheet for guidance.

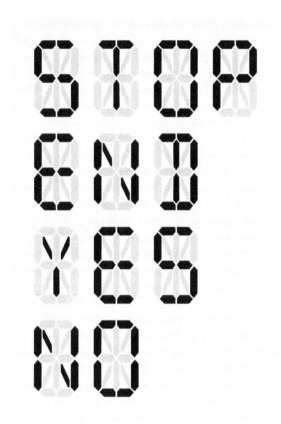

Figure 17-8. *The same text messages shown previously using 7-segment LCDs are shown here using 16-segment displays.*

Dot-matrix LCDs are usually packaged in arrays consisting of eight or more columns and two or more rows of characters. The number of columns is always stated before the number of rows, so that a typical 8 x 2 display contains eight alphanumeric characters in two horizontal rows. An array of characters is properly referred to as a *display module*, but may be described, confusingly, as a *display*, even though a single seven-segment LCD is itself a display. A 16x2 display module is shown from the front in Figure 17-11 and from the rear in Figure 17-12.

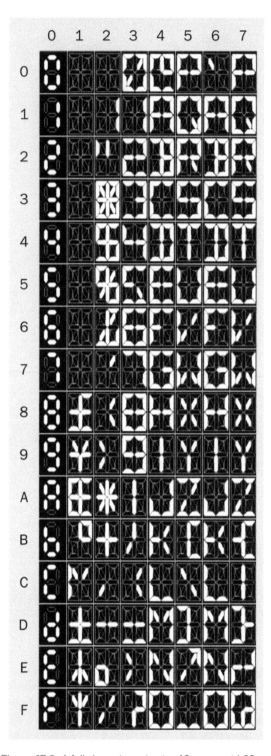

Figure 17-9. *A full character set using 16-segment LCDs.*

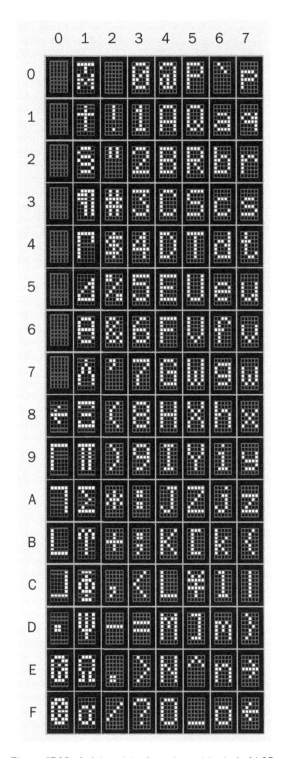

Figure 17-10. *A dot-matrix character set typical of LCDs capable of displaying a matrix of 5×7 dots.*

Figure 17-11. *A 16x2 LCD display module seen from the front.*

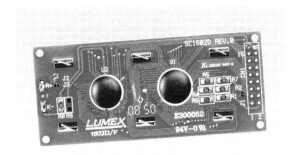

Figure 17-12. *The same 16x2 LCD display module from the previous figure, seen from the rear.*

Multiple-character display modules have been widely used in consumer electronics products such as audio components and automobiles where simple status messages and prompts are necessary—for example, to show the volume setting or broadcast frequency on a stereo receiver. Backlighting is almost always used.

Because the cost of small, full-color, high-resolution LCD screens has been driven down rapidly by the mass production of cellular phones, color displays are likely to displace monochrome dot-matrix LCD display modules in many applications. Similarly, touchscreens will tend to displace pushbuttons and tactile switches. Touchscreens are outside the scope of this encyclopedia.

Color

The addition of filters to create a full color display is shown in simplified form in Figure 17-13.

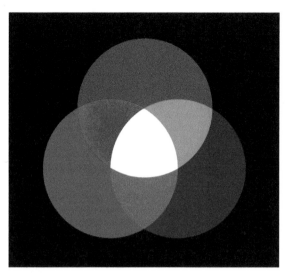

Figure 17-14. *When colors red, green, and blue are transmitted directly to the eye, pairs of these additive primaries create secondary colors cyan, magenta, and yellow. Combining all three additive primaries creates an approximation of white light. This can be verified by viewing a color monitor with a magnifying glass.*

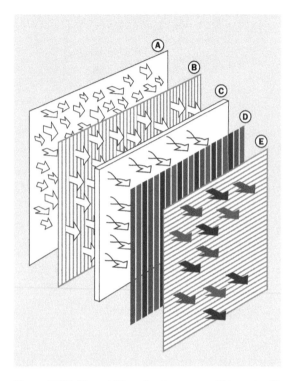

Figure 17-13. *The addition of red, green, and blue color filters, in conjunction with variable density liquid crystal pixels, enables an LCD full-color display.*

Red, green, and blue are almost always used as primary colors for transmitted light, because the combination of different intensities of these *RGB* primaries can create the appearance of many colors throughout the visible spectrum. They are said to be *additive* primaries, as they create brighter colors when they are combined. The principle is illustrated in Figure 17-14.

The use of the word "primaries" to refer to red, green, and blue can cause confusion, as full-color printed materials use a different set of *reflective* primaries, typically cyan, magenta, and yellow, often with the addition of black. In this *CMYK* system, additional layers of pigment will absorb, or subtract, more visible frequences. See Figure 17-15.

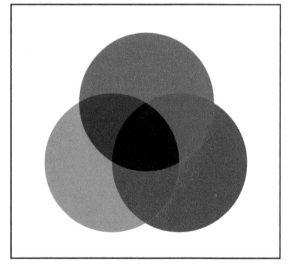

Figure 17-15. *When ink colors cyan, magenta, and yellow are superimposed on white paper and are viewed in white light, pairs of these subtractive primaries create secondary colors red, green, and blue. Overprinting all three subtractive primaries creates an approximation of black, limited by the reflective properties of available pigments. Black ink is usually added to provide additional contrast.*

The complete range of colors that can be created as a combination of primaries is known as the *gamut*. Many different RGB color standards have been developed, the two most widely used being *sRGB* (almost universal in web applications) and *Adobe 1998* (introduced by Adobe Systems for Photoshop, providing a wider gamut). None of the available systems for color reproduction comes close to creating the full gamut that can be perceived by the human eye.

Backlighting Options

For monochrome LCDs, **electroluminescent** backlighting may be used. It requires very low current, generates very little heat, and has a uniform output. However, its brightness is severely limited, and it requires an inverter that adds significantly to the current consumption.

For full-color LCDs, **fluorescent** lights were originally used. They have a long lifetime, generate little heat, and have low power consumption. However, they require a relatively high voltage, and do not work well at low temperatures. Early flat screens for laptop computers and desktop monitors used *cold-cathode fluorescent panels.*

Subsequently, white light-emitting diodes (**LED**s) were refined to the point where they generated a range of frequencies that was considered acceptable. Light from the LEDs passes through a diffuser to provide reasonably consistent illumination across the entire screen. LEDs are cheaper than fluorescent panels, and allow a thinner screen.

High-end video monitors use individual red, green, and blue LEDs instead of a white backlight. This eliminates the need for colored filters and produces a wider gamut. So-called *RGB* LCD monitors are more expensive but are preferred for professional applications in video and print media where accurate color reproduction is essential.

Zero-Power Displays

Some techniques exist to create LCDs that require power only to flip them to and fro between their transparent and opaque states. These are also known as *bistable* displays, but have not become as widely used. They are similar in concept to *e-ink* or *electronic paper* displays, but the principle of operation is different.

How to Use It

So long as an LCD consists of just one numeral, it can be driven by just one decoder chip that translates a binary-coded input into the outputs required to activate the appropriate segments of the LCD. The evolution of multi-digit displays, alphanumeric displays, dot-matrix displays, and graphical displays has complicated this situation.

Numeric Display Modules

An LCD consisting of a single digit is now a rare item, as few circuits require only one numeral for output. More commonly, two to eight numerals are mounted together in a small rectangular panel, three or four numerals being most common. A typical digital alarm clock uses a four-digit numeric display module, incorporating a colon and indicators showing AM/PM and alarm on/off. Other numeric display modules may include a minus sign.

Modules that are described as having 3.5 or 4.5 digits contain three full digits preceded by a numeral 1 composed of two segments. Thus, a 3-digit module can display numbers from 000 through 999, while a 3.5-digit display can display numbers from 000 through 1999, approximately doubling the range.

Numeric display modules of the type described here do not contain any decoder logic or drivers. An external device, such as a microcontroller, must contain a lookup table to translate a numeric value into outputs that will activate the appropriate segments in the numbers in a display, with or without decimal points and a minus sign. To avoid reinventing the wheel, a programmer may download code libraries for microcontrollers to drive commonly used numeric display modules. It is important to remember, though, that segments in monochrome LCDs must be activated by AC, typically a square wave with a frequency of 30Hz to 90Hz.

An alternative is to use a decoder chip such as the 4543B or 4056B, which receives a binary-coded decimal input (i.e., 0000 through 1001 binary, on four input puts) and translates it into an output on seven pins suitable for connection with the seven segments of a 7-segment display. The 4543B requires a square-wave input to its "phase" pin. The square-wave must also be applied simultaneously to the backplane of the LCD, often identified as the "common" pin on datasheets. Pinouts for the 4543B are shown in Figure 17-16.

The 4543B includes provision for "display blanking," which can be used to suppress leading zeros in a multidigit number. However, the lack of outputs to control a minus sign or decimal point limits the decoder to displaying positive integers.

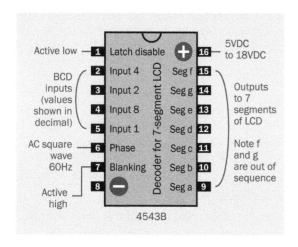

Figure 17-16. *Pinouts for the 4543B decoder chip, which is designed to drive a seven-segment numeric LCD.*

The power supply for a 4543B can range from 5VDC to 18VDC, but because the logic-high output voltage will be almost the same as that of the power supply, it must be chosen to match the power requirements of the LCD (very often 5VAC).(((

To drive a three-digit numeric display module, a separate decoder chip can be used to control each digit. The disadvantage of this system is that each decoder requires three inputs, so that a three-digit display will require nine outputs from the microcontroller.

To deal with this issue, it is common to *multiplex* a multi-digit display. This means that each output from the decoder is shared among the same segments of all the LCD numerals. Each LCD numeral is then activated in sequence by applying AC voltage to its common pin. Simultaneously, the decoder sends the data appropriate to that LCD. This process must be fast enough so that all the digits appear to be active simultaneously, and is best managed with a microcontroller. A simplified schematic is shown in Figure 17-17. It can be compared with a similar circuit to drive LED displays, shown in Figure 24-13.

Alphanumeric Display Module

Arrays of dot-matrix LCDs that can display alphabetical characters as well as numerals require preset character patterns (usually stored in ROM) and a command interpreter to process instructions that are embedded in the data stream. These capabilities are often built into the LCD module itself.

While there is no formal or de facto standard, the command set used by the Hitachi HD44780 controller is installed in many displays, and code libraries for this set are available for download from sites dedicated to the Arduino and other microcontrollers. Writing code from scratch to control all aspects of an alphanumeric display is not a trivial chore. The Hamtronix HDM08216L-3-L30S is a display that incorporates the HD44780.

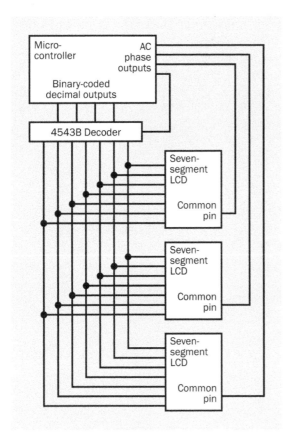

Figure 17-17. *When two or more numeric displays are multiplexed, a control device (typically, a microcontroller) activates each of them in turn via its backplane (common terminal) while sending appropriate data over a shared bus.*

Regardless of which standard is used, some features of alphanumeric display modules are almost universal:

- Register select pin. Tells the display whether the incoming data is an instruction, or a code identifying a displayable character.
- Read/write pin. Tells the display whether to receive characters from a microcontroller or send them to a microcontroller.
- Enable/disable pin.
- Character data input pins. There will be eight pins to receive the 8-bit ASCII code for each displayable character in parallel. Often there is an option to use only four of these pins, to

reduce the number of microcontroller outputs necessary to drive the display. Where four pins are used, each 8-bit character is sent in two segments.

- LED backlight pin. Two may be provided, one connected to the anode(s) of the LED backlight, the other to the cathode(s).
- Reset pin.

Embedded instruction codes can be complex, including commands to reposition the cursor at a specific screen location, backspace-and-erase, scroll the display, and erase all characters on the screen. Codes may be included to adjust screen brightness and to switch the display between light-on-dark (negative) and dark-on-light (positive) characters.

Some display modules also have graphics capability, allowing the user to address any individual pixel on the screen.

Because of the lack of standardization in control codes, manufacturer's datasheets must be consulted to learn the usage of a particular alphanumeric display module. In addition to datasheets, online user forums are a valuable source of information regarding quirks and undocumented features.

What Can Go Wrong

Temperature Sensitivity

Liquid crystals vary in their tolerance for low and high temperatures, but generally speaking, a higher voltage may be necessary to create a sufficiently dense image at a low temperature. Conversely, a lower voltage may be necessary to avoid "ghosting" at a high temperature. An absolutely safe operating temperature range is likely to be 0 through 50 degrees Celsius, but check the manufacturer's datasheet for confirmation. Special-purpose LCDs are available for extreme temperatures.

Excessive Multiplexing

A twisted nematic display is likely to perform poorly if its duty cycle is greater than 1:4. In other words, more than four displays should not be multiplexed by the same controller.

DC Damage

An LCD can be damaged quickly and permanently if it is subjected to DC current. This can occur by accident if, for example, a timer chip is being used to generate the AC pulse stream, and the timer is accidentally disconnected, or has an incorrect connection in its RC network. Check timer output with a meter set to measure AC volts before allowing any connection to the common pin of an LCD.

Bad Communications Protocol

Many alphanumeric display modules do not use a formal communications protocol. Duplex serial or I2C connection may not be available. Care must be taken to allow pauses of a few milliseconds after execution of embedded commands, to give the display sufficient time to complete the instruction. This is especially likely where a command to clear all characters from the screen has to be executed. If garbage characters appear on the screen, incorrect data transfer speed or lack of pause times may be to blame.

Wiring Errors

This is often cited by manufacturers as the most common cause of failure to display characters correctly, or lack of any screen image at all.

incandescent lamp 18

The terms *incandescent light*, *incandescent bulb*, and *incandescent light bulb* are often used interchangeably with **incandescent lamp**. Because the term "lamp" seems to be most common, it is used here. A *panel-mounted indicator lamp* is considered to be an assembly containing an incandescent lamp.

A *carbon arc*, which generates light as a self-sustaining spark between two carbon electrodes, can be thought of as a form of incandescent lamp, but is now rare and is not included in this encyclopedia.

OTHER RELATED COMPONENTS

- **LED area lighting** (see Chapter 23)
- **LED indicator** (see Chapter 22)
- **neon bulb** (see Chapter 19)
- **fluorescent light** (see Chapter 20)

What It Does

The term *incandescent* describes an object that emits visible light purely as a consequence of being hot. This principle is used in an **incandescent lamp** where a wire *filament* glows as a result of electric current passing through it and raising it to a high temperature. To prevent oxidation of the filament, it is contained within a sealed bulb or tube containing an inert gas under low pressure or (less often) a vacuum.

Because incandescent lamps are relatively inefficient, they are not considered a wise environmental choice for area lighting and have been prohibited for that purpose in some areas. However, small, low-voltage, panel-mount versions are still widely available. For a summary of advantages of miniature incandescent lamps relative to **light-emitting diodes** (LEDs) see "Relative Advantages" on page 179.

Schematic symbols representing an incandescent lamp are shown in Figure 18-1. The symbols are all functionally identical except that the one at bottom right is more likely to be used to represent small panel-mounted indicators.

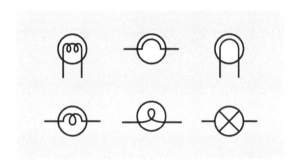

Figure 18-1. *A variety of symbols can represent an incandescent lamp. The one at bottom right may be more commonly used for small panel-mounted indicators.*

The parts of a generic incandescent light bulb are identified in Figure 18-2:

A: Glass bulb.

B: Inert gas at low pressure.

171

C: Tungsten filament.

D: Contact wires (connecting internally with brass base and center contact, below).

E: Wires to support the filament.

F: Internal glass stem.

G: Brass base or cap.

H: Vitreous insulation.

I: Center contact.

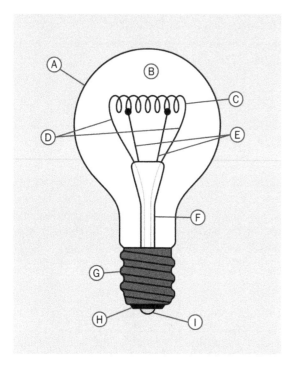

Figure 18-2. *The parts of a typical incandescent lamp (see text for details).*

History

The concept of generating light by using electricity to heat a metal originated with Englishman Humphrey Davy, who demonstrated it with a large battery and a strip of platinum in 1802. Platinum was thought to be suitable because it has a relatively high melting point. The lamp worked but was not practical, being insufficiently bright and having a short lifespan. In addition, the platinum was prohibitively expensive.

The first patent for an incandescent lamp was issued in England in 1841, but it still used platinum. Subsequently, British physicist and chemist Joseph Swan spent many years attempting to develop practical carbon filaments, and obtained a patent in 1880 for *parchmentized thread*. His house was the first in the world to be illuminated by light bulbs.

Thomas Edison began work to refine the electric lamp in 1878, and achieved a successful test with a carbonized filament in October 1879. The bulb lasted slightly more than 13 hours. Lawsuits over patent rights ensued. Carbonized filaments were used until a tungsten filament was patented in 1904 by the German/Hungarian inventor Just Sándor Frigyes and the Croatian inventor Franjo Hanaman. This type of bulb was filled with an inert gas, instead of using a vacuum.

Many other pioneers participated in the effort to develop electric light on a practical basis. Thus it is incorrect to state that "Thomas Edison invented the light bulb." The device went through a very lengthy process of gradual refinement, and one of Edison's most significant achievements was the development of a power distribution system that could run multiple lamps in parallel, using filaments that had a relatively high resistance. His error was insisting on using *direct current* (DC) while his rival Westinghouse pioneered *alternating currrent* (AC), enabling power transmission over longer distances through the use of **transformers**. The use of AC also enabled Tesla's brushless induction motor.

By the mid-1900s, most incandescent bulbs used tungsten filaments.

How It Works

All objects emit electromagnetic radiation as a function of their temperature. This is known as *black body radiation*, based on the concept of an object that absorbs all incoming light, and thus does not reflect any sources from outside itself. As its temperature increases, the intensity of the radiation increases while the wavelength of the radiation tends to decrease.

If the temperature is high enough, the wavelength of the radiation enters the visible spectrum, between 380 and 740 nanometers. (A nanometer is one-billionth of a meter.)

The melting point of tungsten is 3,442 degrees Celsius, but a lamp filament typically operates between 2,000 and 3,000 degrees. At the higher end of this scale, evaporation of metal from the filament tends to cause deposition of a dark residue on the inside of the bulb, and erodes the filament more rapidly, to the point where it eventually breaks. At the lower end of this scale, the light will be yellow and the intensity will be reduced.

Spectrum

The color of black-body radiation is measured using the Kelvin temperature scale. The increment of 1 degree Kelvin is the same as 1 degree Celsius, but the Kelvin scale has a zero value at *absolute zero*. This is the theoretical lowest conceivable temperature, at which there is complete absence of heat. It is approximately –273 degrees Celsius.

From this it is evident that if K is a temperature in degrees Kelvin and C is a temperature in degrees Celsius:

K = C + 273 (approximately)

Calibration of light sources in degrees Kelvin is common in photography. Many digital cameras allow the user to specify the *color temperature* of lights that are illuminating an indoor scene, and the camera will compensate so that the light source appears to be pure white with all colors in the visible spectrum being represented equally.

Some computer monitors also allow the user to specify a white value in degrees Kelvin.

Color temperature is used in astronomy, because the spectrum of many stars is comparable with that of a theoretical black body.

A color temperature of 1,000 degrees K will have a dark orange hue, while 15,000 degrees K or higher will have a blue hue comparable to that of a pale blue sky. The color temperature of the sun is approximately 5,800 K. Interior lighting is often around 3,000 K, which many people find acceptable because it creates pleasant flesh tones. An incandescent bulb described by the manufacturer as "soft white" or "warm" will have a lower color temperature than one which is sold as "pure white" or "paper white."

Graphs showing the emission of wavelengths at various color temperatures are shown in Figure 18-3. The rainbow section indicates the approximate range of visible wavelengths between ultraviolet, on the left, and infrared, on the right. For purposes of clarity, the peak intensity for each color temperature has been equalized. In reality, increasing the temperature also increases the light output.

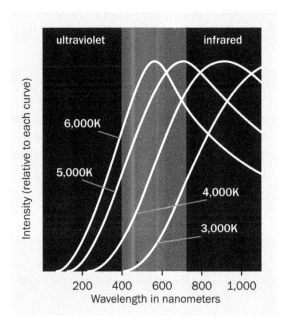

Figure 18-3. *Approximate peak wavelengths for blackbody radiation at various color temperatures in degrees Kelvin. The curves have been adjusted so that their peak values are equalized. Adapted from an illustration in the reference book Light Emitting Diodes by E. Fred Schubert.*

Non-Incandescent Sources

So long as light is generated by heating a filament, plotting the intensity against wavelength will result in a smooth curve without irregularities. A higher Kelvin value will simply displace and compress the curve laterally without changing its basic shape to a significant degree.

The introduction of **fluorescent** sources and, subsequently, light-emitting diodes (**LEDs**) has complicated this scenario. Because they are *luminescent* rather than *incandescent*, they do not generate an evenly weighted, continuous range of wavelengths.

LEDs tend to emit monochromatic light, meaning that it is tightly centered around just one color. A "white" LED is really a blue LED in which a phosphor coating on the semiconductor die is excited to create light over a broader range. A fluorescent light tends to create *spectral lines* which show up as sharp peaks at a few wavelengths determined by the mercury inside the bulb. Figure 18-4 illustrates these problems.

The human eye tends to compensate for the yellow emphasis of incandescent lamps and for the irregularities in spectra emitted by other light sources. Also, the eye is often unable to distinguish between "white" light created as a mix of all the visible wavelengths, and light that appears white even though it is dominated by a few isolated wavelengths from a fluorescent source.

However, when the eye views colors that are illuminated by a source that has gaps in its spectrum, some of the colors will appear unnaturally dull or dark. This is true also if an imperfect source is used as a backlight to create colors on a video monitor. Colors rendered by different light sources are shown in Figure 23-7 and subsequent figures.

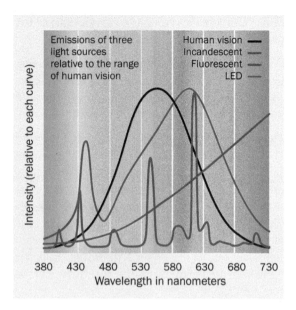

Figure 18-4. *The relative performance of three light sources compared with sensitivity of the human eye to the visible spectrum. Note that the range of wavelengths on the horizontal scale in this figure is not the same as the range in the previous figure. The color assigned to each curve is arbitrary. Adapted from VU1 Corporation.*

Photography is adversely affected by the use of LEDs or fluorescents as a light source. Reds, for example, can seem dark when lit by white LEDs, while blues can be inappropriately intense. Because the source does not have an emission curve comparable to that of an incandescent light, the auto-white balance feature of a digital camera may be unable to address this problem, and it cannot be resolved by entering a different Kelvin number manually.

The fidelity with which a light source is capable of displaying the full visible spectrum is known as the *color rendering index* (CRI), ranging from a perfect score of 100 down to 0 or even lower (sodium-vapor street lighting has a negative value). Computing the index requires standard reference color samples and has been criticized for generating scores that do not correlate well with subjective assessments.

Incandescent bulbs can have a CRI of 100, while an uncorrected "white" LED may score as low as 80.

Power Consumption

Approximately 95% of the power consumed by an incandescent lamp generates heat instead of visible light. This wastage of power in room lighting is compounded by the power consumption of air conditioning to remove the heat from enclosed spaces in hot climates. While the heat from incandescent lamps does reduce the need for space heating in cold environments, heat is delivered more efficiently by using systems designed for that purpose. Consequently, greater energy efficiency can be achieved with a light source that generates less heat, regardless of ambient air temperature.

Variants

Miniature Lamps

Prior to the development of **LED**s, all light-emitting panel-mounted indicators were either **neon bulbs** or **incandescent lamps**. The use of neon is limited by its need for a relatively high voltage.

Miniature incandescents were the traditional choice for battery-powered light sources, and at the time of writing are still used in cheap flashlights. Variants are available that are as small as a 5mm LED, with a claimed life expectancy that is comparable, although they draw more current to generate an equivalent light intensity, because much of their power is wasted in infrared wavelengths.

The photograph in Figure 18-5 is of a miniature lamp terminating in pins spaced 0.05" apart. The total height of the lamp, including its ceramic base, is less than 0.4," while its diameter is just over 0.1". It draws 60mA at 5V and is rated for 25,000 hours.

The photograph in Figure 18-6 is of a lamp of similar size and power consumption, but terminating in wire leads and rated for 100,000 hours. It emits 0.63 lumens.

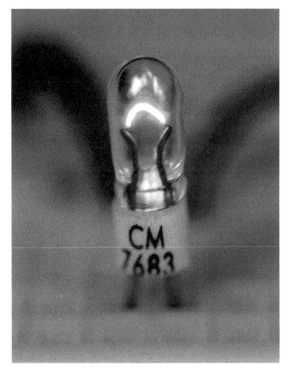

Figure 18-5. *A miniature lamp less than 0.4" high, terminating in pins spaced 0.05" apart.*

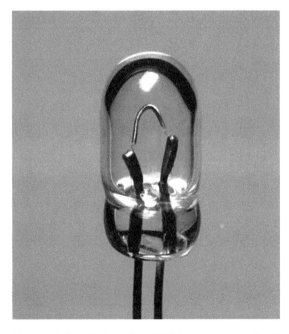

Figure 18-6. *This lamp is 0.25" high and terminates in wire leads.*

The lamp pictured in Figure 18-7 is slightly larger, with a glass envelope about 0.25" diameter. It is rated for less than half the lifetime of the lamp in Figure 18-6 but emits three times as much light —a typical tradeoff. Various base styles are available.

Figure 18-7. This lamp has a glass envelope about 0.35" high. Its screw-in base makes it easier to replace than an LED.

In the United States, the light output from miniature incandescent lamps may be measured in lumens, but is more often rated in *mean spherical candlepower* (MSCP). An explanation of light measurement is included in "MSCP" on page 178.

Lamp lenses provide a quick and simple way to add color to a miniature incandescent lamp. Usually the lens is cylindrical with a hemispherical end cap, and is designed to push-fit or snap-fit over a small lamp. Even when the cap is translucent, it may still be referred to as a lens.

Panel-Mount Indicator Lamps

This term often refers to a tubular assembly containing a miniature lamp, ready for installation.

The enclosure is often designed to snap-fit into a hole drilled in the panel. If the incandescent bulb inside the enclosure cannot be replaced, the component is said to be "non-relampable." Figure 18-8 shows a 12-volt panel-mount indicator lamp.

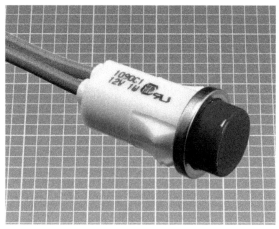

Figure 18-8. This panel-mount indicator lamp is designed to push-fit into a hole 1/2" in diameter. The bulb inside it is not replaceable, causing the assembly to be classified as "non-relampable."

Halogen or Quartz-Halogen

This is a type of incandescent lamp containing gases under pressure in which halogens such as iodine or bromine cause evaporated tungsten atoms to be redeposited on the filament. A halogen lamp can therefore operate at a higher temperature, creating a light that is less yellow and brighter than that from a comparable incandescent lamp. It also enables a smaller bulb, but requires an envelope of borosilicate-halide glass (often termed *fused quartz*) instead of regular glass. A halogen lamp will be slightly more efficient than an incandescent bulb of the same wattage, and will last longer.

Halogens are available in a variety of formats. The small bulb pictured in Figure 18-9 consumes 75W, emitting 1,500 lumens at 3,000 degrees Kelvin. The light intensity is claimed to be equivalent to that of a 100W incandescent bulb. It has a mini-candelabra base.

Oven Lamps

Oven lamps are designed to withstand the high temperature in an oven. Typically they are usable with ambient temperatures up to 300 degrees C. A common power rating is 15W.

Figure 18-9. *A halogen bulb slightly more than 2" in length, designed for 115VAC.*

Base Variants

Miniature lamps are available with a wide variety of connection options, including wire terminals, single-contact bayonet, double-contact bayonet, miniature screw base, and fuse style. Most of these options require a matching socket.

Screw-in lamps for room illumination are common in household lighting in the United States and many other countries (but not in the UK, where bayonet fittings are used). The US socket size is designated by letter E followed by a number that gives the socket diameter in millimeters. Common sizes are E10, E14 and E27.

A *bayonet base* is fitted with two small lugs protruding on opposite sides. The lamp is secured by pushing it in and twisting it to engage the lugs in slots in the socket. The advantage of a bayonet

base is that the bulb is less likely to become loose as a result of vibration.

A *pin base* consists simply of a pair of pins that will push-fit into small holes in a socket.

A *flange base* has a flange that engages in a socket where flexible segments will retain it.

A *wedge base* is forced between two contacts which retain the bulb by friction.

Some indicator lamps terminate simply in long, thin leads that can be soldered.

Values

While the power consumption of full-size incandescent lamps is rated in watts, small indicator lamps are rated in milliamps at the voltage for which they are designed. Miniature lamps may require specific voltages ranging from as low as 2V to 24V. A higher voltage generally necessitates a longer filament, which may entail a larger bulb.

The light that a lamp will emit can be measured in two ways: either as the power of the lamp (not its power consumption, but its radiating power), or as the light delivered to a specific area at a specific distance. These two measurements may differ because a lamp may concentrate its light in a beam, as in the case of a reflector bulb or an LED.

Power

Flux, in watts, is a measurement of energy flow in joules per second. The total radiating power of a lamp, in all wavelengths, in all directions, is known as its *radiant flux*. Because invisible wavelengths are of little interest when assessing the brightness of a lamp, the term *luminous flux* is used to describe the apparent brightness of the lamp in the visible spectrum. The unit for luminous flux is the *lumen*.

The human eye is most responsive to yellow-green hues in the center of the spectrum. Consequently, the measurement of luminous flux is weighted toward green at a wavelength of 555 nanometers. Red and violet are considered to have low luminous flux, while infrared and ultraviolet have a zero value.

When considering a value expressed in lumens, remember:

- Lumens are a measure of the total radiated power output of a light source, in all directions, in the visible spectrum only, weighted toward the characteristics of the human eye.

- The number of lumens of a light source does not define the direction in which the light is shining, or its uniformity.

- The abbreviation for lumen is *lm*.

A conventional incandescent lamp that consumes 100W of electricity is likely to have a light output of about 1,500 lumens. A 40W fluorescent tube can have a light output of about 2,600 lumens.

Illuminance

The *illuminance* of a light source is defined as the luminous flux per unit of area. This can be thought of as the brightness of a surface illuminated by the source.

Illuminance is measured in *lux*, where 1 lux = 1 lumen per square meter. For accurate calibration, the illuminated surface should be spherical in shape, and must be located 1 meter from the light source, with the source at the geometrical center of the sphere.

Illuminance used to be measured in *foot-candles*, where 1 foot-candle was 1 lumen per square foot.

- The number of lumens per square meter (lux) does not define the size of the illuminated area, only the brightness per unit of area.

- A lamp that has a tightly focused beam can achieve a high lux rating. When selecting a lamp for an application, the angle of dispersion of the beam must be considered in conjunction with its lux rating.

Intensity

A *candela* measures the luminous flux within an angle of dispersion. The angle is three-dimensional, and can be imagined as the sharpness of a point of a cone, where the light source is at the point and the cone represents the dispersion of light.

The three-dimensional angle of dispersion is measured in *steradians*. If a light source is at the center of a sphere that has a radius of 1 meter, and is illuminating one square meter of the surface of the sphere, the angle of dispersion is 1 steradian.

- A source of 1 lumen which projects all its light through a dispersion angle of 1 steradian is rated at 1 candela.

- The number of candelas does not define the angle of dispersion, only the intensity within that angle.

- A light source rated for 1,000 candelas could have a power of 10 lumens concentrated within an angle of 0.01 steradians, or could have a power of only 1 lumen concentrated within an angle of 0.001 steradians.

- There are 1,000 millicandelas in 1 candela. The abbreviation for candela is *cd* while the abbreviation for a millicandela is *mcd*.

- LEDs are often rated in mcd. The number describes the intensity of light within its angle of dispersion.

MSCP

Although the term *candlepower* is obsolete, it has been redefined as being equal to 1 candela. *Mean spherical candlepower* (MSCP) is a measurement of all the light emitted from a lamp in all directions. Because the light is assumed to be omnidirectional, it fills 4 * π (about 12.57) steradians. Therefore 1 MSCP = approximately 12.57

lumens. In the United States, MSCP is still the most common method of rating the total light output of a miniature lamp.

Efficacy

The *radiant luminous efficacy* (abbreviated *LER*) assesses how effective a lamp is at channeling its output within the visible spectrum, instead of wasting it in other wavelengths, especially infrared. LER is calculated by dividing the power emitted in the visible spectrum (the *luminous flux*) by the power emitted over all wavelengths.

Thus, if VP is the power emitted in the visible spectrum, and AP is the power emitted in all wavelengths:

```
LER = VP / AP
```

LER is expressed in *lumens per watt*. It can range from a low value of around 12 lm/W for a 40W incandescent bulb to 24 lm/W for a quartz halogen lamp. Fluorescent lamps may average 50 lm/W. LEDs vary, but can achieve 100 lm/W.

Efficiency

The *radiant luminous efficiency* (abbreviated *LFR*) of a lamp measures how good its radiant luminous efficacy is, compared with an imaginary ideal lamp. (Note the difference between the words "efficiency" and "efficacy.") LFR is determined by dividing the radiant luminous efficacy (LER) by the maximum theoretical LER value of 683 lm/W, and multiplying by 100 to express the result as a percentage. Thus:

```
LFR = 100 * ( LER / 683 )
```

The LFR ranges from around 2% for a 40W bulb to 3.5% for a quartz halogen lamp. LEDs may be around 15% while fluorescents are closer to 10%.

How to Use It

When first introduced, LEDs were limited by their higher price, lower maximum light output, and inability to display blue or white. The price difference has disappeared for small indicators, while gaps in the color range have been filled

(although the color rendering index of LEDs is still inferior).

Brightness remains an advantage for large incandescents relative to LEDs, as they are more upwardly scalable. However, fluorescents and vapor lamps have an advantage for very high light output, as in the lighting of big-box stores or parking lots. Thus the range of applications for incandescent bulbs is diminishing, especially because common types are now illegal for domestic light fixtures in many parts of the world.

Relative Advantages

When choosing whether to use an incandescent lamp or an LED, these advantages of an incandescent lamp should be considered:

- The intensity can be adjusted with a **triac**-based dimmer. Regular fluorescents cannot be dimmed, while LEDs often require different dimmer circuitry.

- The intensity can also be adjusted with a rheostat. The output from fluorescents cannot.

- Easy white-balance correction. LEDs and fluorescents do not naturally produce a consistent output over the visible spectrum.

- Can be designed to operate directly from a wide range of voltages (down to around 2V and up to around 300V). A higher voltage entails a longer filament wire, which may require a larger bulb. LEDs require additional components and circuitry to use higher voltages.

- Incandescent bulbs are more tolerant of voltage fluctuations than LEDs. With battery operation, the incandescent will still provide some reduced light output when the voltage has diminished radically. LEDs will not perform at all at currents lower than their threshold.

- An incandescent is nonpolarized and may be socketed, which simplifies user replacement.

LEDs are polarized and are usually soldered in.

- Can be powered by AC or DC without any modification or additional circuitry. LEDs require DC, which must be provided through a transformer and rectifier, or similar electronics, if AC power is the primary source.

- Can be equally visible from a wide range of viewing angles. LEDs have restricted viewing angles.

- The heat output from an incandescent bulb may occasionally be useful (for example in a terrarium, or in incubators for poultry).

- Trouble-free switching. Fluorescents tend to hesitate and blink when power is applied, and they require a *ballast* to energize them. The lifespan of fluorescents is reduced by frequent switching.

- No low-temperature problems. Incandescent lamps are not significantly affected by low temperatures. Fluorescents may not start easily in a cold environment, and may flicker or glow dimly for 10 minutes (or more) until they are warm enough to function properly.

- Easy disposal. Fluorescent lights contain small quantities of mercury that are an environmental hazard. They should not be mixed with ordinary trash. Compact fluorescent lamps (CFLs) and LEDs used for room lighting will be packaged with electronics that should ideally be recycled, although this is not very practical. Incandescent bulbs impose the least burden on the environment when they are thrown away.

However, the incandescent lamp has some obvious disadvantages:

- Relatively inefficient.
- More susceptible to vibration.
- More fragile.
- Likely to have a shorter natural life expectancy than LEDs, fluorescents, or neon bulbs, although the lifetime of a small panel indicator can be equal to that of an LED if a low color temperature is acceptable.

- Requires a filter or tinted glass envelope to generate colored light. This further reduces the lamp's efficiency.

- Cannot be miniaturized to the same degree as an LED indicator.

Derating

The lifespan of a lamp can be greatly extended by choosing one with a higher current rating or using it at a lower voltage. The light output will be reduced, and the color temperature will be at a lower Kelvin number, but in some situations this tradeoff may be acceptable.

The graphs in Figure 18-10 suggest that if the voltage of a hypothetical miniature lamp is reduced to 80% of the manufacturer's recommended value, this can make the lamp last 20 times as long. Note, however, that this will cut the light intensity to 50% of its normal value.

Conversely, using 130% normal voltage will give 250% of the normal light output, while shortening the life of the lamp to 1/20 of its normal value. Naturally these figures are approximations that may not apply precisely to a specific lamp.

What Can Go Wrong

High Temperature Environment

If an incandescent lamp is used in an environment hotter than 100 degrees Celsius, the life of the lamp is likely to be reduced by the "water cycle." Any water molecules inside the glass envelope will break down, allowing oxygen to combine with the tungsten filament to form tungsten oxide. The tungsten is deposited on the inside of the glass while the oxygen is liberated and begins a new cycle.

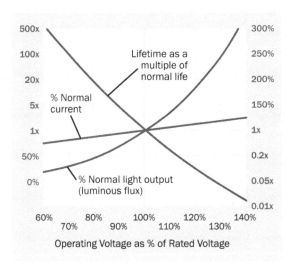

Figure 18-10. *The life expectancy of a hypothetical minia-ture lamp is very strongly influenced by voltage. Applying only 60% of the rated voltage can make a lamp last 500 times its normal lifespan, although it will greatly reduce light output. Note that the vertical axes apply to curves of the same color. Adapted from "Characteristics of Minia-ture Lamps" from Toshiba Lighting and Technology Cor-poration.*

Fire Risk

The partially evacuated bulb of an incandescent lamp provides some separation and protection from the heat in the filament, but if the bulb can-not disperse heat by radiation or convection, its temperature can rise to the point where it ignites flammable materials.

Halogen lamps have an elevated fire risk because they operate at a higher temperature and are smaller, providing less surface area to disperse the heat. They also contain gases under seven to eight atmospheres of pressure. Thermal stress can cause a halogen bulb to shatter, and finger-prints on the glass can increase this risk.

Current Inrush

When an incandescent lamp is first switched on, its filament has one-tenth the resistance that it will exhibit when it becomes hot. Consequently, the lamp will take a large initial surge of current, which stabilizes after about 50 milliseconds. This should be considered if one or more small lamps shares a DC power supply with components such as logic chips that may be sensitive to voltage fluctuations.

Replacement Problems

Because of the limited life of incandescent lamps, they should be installed in such a way that they are easy to replace. This can be an issue with panel indicators, where disassembly of a device may be necessary to reach the lamp.

The range of small incandescent lamps is dimin-ishing, and may continue to diminish in the fu-ture. Future availability of replacement lamps should be considered when designing a circuit. When building equipment in small quantities, spare lamps should be purchased for future use.

neon bulb | 19

The terms **neon bulb**, *neon indicator*, and *neon lamp* tend to be used interchangeably. In this encyclopedia, a neon *bulb* is defined as a glass capsule containing two electrodes in neon gas (or a combination of gases in which neon is present). A neon *lamp* is an assembly containing a neon bulb, usually using a plastic tube with a tinted transparent cap at one end. A neon *indicator* is a miniature neon lamp that is usually panel-mounted.

Large-scale neon tubes used in signage are not included in this encyclopedia.

OTHER RELATED COMPONENTS

- **incandescent lamp** (see Chapter 18)
- **fluorescent light** (see Chapter 20)
- **LED indicator** (see Chapter 22)

What It Does

When voltage is applied between two electrodes inside a neon bulb, the inert gas inside the bulb emits a soft red or orange glow. This color may be modified by using a tinted transparent plastic cap, known as a *lens*, in a *neon lamp* assembly.

A neon bulb is usually designed for a power supply of 110V or higher. It functions equally well with alternating or direct current.

The schematic symbols in Figure 19-1 are commonly used to represent either a neon bulb or a neon lamp. They are all functionally identical. The black dot that appears inside two of the symbols indicates that the component is gas filled. The position of the dot inside the circle is arbitrary. Even though all neon bulbs are gas filled, the dot is often omitted.

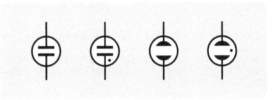

Figure 19-1. *Any of these symbols may represent a neon bulb or a neon lamp. The dot in two of the symbols indicates that the component is gas filled. All neon bulbs are gas filled, but the dot is often omitted.*

The photograph in Figure 19-2 shows a neon bulb with a series resistor preattached to one lead. Many bulbs are sold in this configuration, because a resistor must be used to limit current through the bulb. The bulb has no polarity and can be used on an AC or DC power supply. The same bulb is shown in its energized state in Figure 19-3.

Figure 19-2. *A typical neon bulb with series resistor attached to one lead.*

top end of the glass tube is heated until it melts, and is pinched off. This creates a distinctive protrusion known as the *pip*.

Figure 19-3. *The same bulb from the previous photograph, energized with 115VAC.*

How It Works

Construction

The parts of a neon bulb are illustrated in Figure 19-4. When the bulb is fabricated, it begins as a glass tube. The leads are made of *dumet*, consisting of a copper sheath around a nickel iron core. This has the same coefficient of expansion as glass, so that when the glass is heated and melted around the leads, it forms a seal that should be unaffected by subsequent temperature fluctuations. This area is known as the *pinch* in the tube.

Nickel electrodes are welded onto the leads before the leads are inserted into the tube. The electrodes have an emissive coating that reduces the minimum operating voltage. The glass tube is filled with a combination of neon and argon gases, or pure neon for higher light output (which reduces the life of the component). The

Ionization

When a voltage is applied between the leads to the bulb, the gas becomes ionized, and electrons and ions are accelerated by the electric field. When they hit other atoms, these are ionized as well, maintaining the ionization level. Atoms are excited by collisions, moving their electrons to higher energy levels. When an electron returns from a higher level to a ground state, a photon is emitted.

This process begins at the *starting voltage* (also known as the *striking voltage*, the *ignition voltage*, or the *breakdown voltage*) usually between 45V and 65V for standard types of bulb, or between 70V and 95V for high-brightness types.

When the bulb is operating, it emits a soft radiance known as a *glow discharge* with a wavelength ranging from 600 to 700 nanometers.

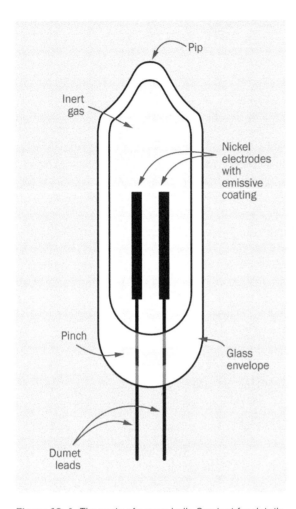

Figure 19-4. *The parts of a neon bulb. See text for details.*

The ionization of the gas allows current to flow through it. This will continue even if the power supply is reduced by 10 to 20 volts to a level known as the *maintaining voltage*.

Negative Resistance

When the glow discharge persists below the starting voltage, this is a form of *hysteresis*, meaning that the neon bulb tends to "stick" in its on state. It remains on while its power supply decreases to the maintaining voltage, but once it

switches off, it will "stick" in its off state until the power supply increases again above the maintaining voltage to the starting voltage. The concept of hysteresis is discussed in the entry on **comparators**. See Figure 6-2.

A neon bulb is said to have *negative resistance*. If the current is allowed to increase without restraint, the resistance eventually decreases while the current increases further. If this runaway behavior is not controlled, the bulb will destroy itself.

This behavior is characteristic of gas-discharge tubes generally. A graph showing this appears in Figure 19-5. Note that both scales are logarithmic. Also note that the curve shows how current will be measured in response to voltage. If the voltage is reduced after it has increased, the transitional events shown by the graph will not recur in reverse order. This is especially true if arcing is allowed to begin, as it will almost certainly destroy the component.

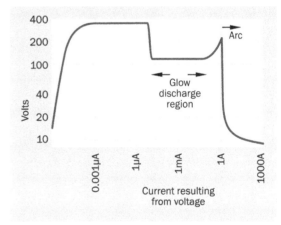

Figure 19-5. *A gas discharge tube, such as a neon bulb, is said to have a negative resistance, as current passing through it tends to increase uncontrollably after the gas is ionized and becomes conductive. (Derived from measurements made by David Knight, on a web page named after his radio ham call sign, G3YNH.)*

A neon bulb can be controlled very simply with a series resistor that maintains it in gas-discharge mode. To understand the operation of the resistor, consider the combination of the lamp and the resistor as a voltage divider, as shown in Figure 19-6. Before the lamp begins to pass current, it has an almost infinite resistance. Therefore, the voltage on both sides of the resistor will be approximately equal, the bulb passes almost no current, and it remains dark.

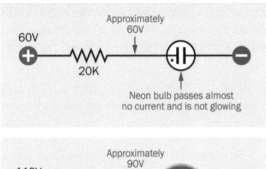

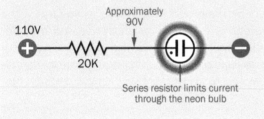

Figure 19-6. *A series resistor is essential to limit the current through a neon bulb.*

After the lamp begins to pass current, the requirement now is for the series resistor to reduce the voltage from the supply level (probably around 110V) to the maintaining level (probably around 90V). This means that the desired voltage drop is 20V, and if the manufacturer's specification tells us that the lamp should pass 1mA (i.e., 0.001 amps), R, the value of the series resistor, is given by Ohm's Law:

    R = 20 / 0.001

Thus, the value for R is 20K. In fact, the value of a resistor supplied with a neon bulb may range from 10K to 220K, depending on the characteristics of the bulb and the supply voltage that will be used.

Now if the bulb's effective internal resistance falls radically, the resistor still limits the currrent. In a hypothetical worst-case scenario, if the bulb's resistance drops all the way to zero, the resistor must now impose the full voltage drop of 110V, and the current, I, will be found by Ohm's law:

    I = 110 / 20,000

That is, about 5mA, or 0.005A.

Neon tubes used in signage require a more sophisticated voltage control circuit which is not included in this encyclopedia.

How to Use It

The use of a neon bulb for an indicator lamp is primarily limited to situations where domestic supply voltage (115VAC or 220VAC) is readily available. "Power on" lights are the obvious application, especially as neon indicators can accept AC. The switch shown in Figure 19-7 is illuminated by an internal neon bulb. The rectangular indicator in Figure 19-8 is designed to run on domestic supply voltage, and its internal bulb and resistor can be clearly seen through the green plastic. The assembly in Figure 19-9 is about 0.5" in diameter, which is the lower limit for neon indicators.

Figure 19-7. *This power switch is illuminated by an internal neon bulb.*

Figure 19-8. *The neon bulb and its series resistor are visible inside this indicator.*

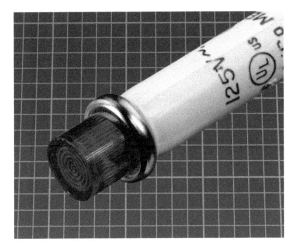

Figure 19-9. *A relatively small neon indicator lamp, designed for insertion in a hole 0.5" diameter.*

Limited Light Output

Neon bulbs have a light output of around 0.06 lumens per milliamp of consumed power (standard brightness type) or 0.15 lumens per milliamp of consumed power (high brightness type).

Comparing this value with the intensity of **LED** indicators is difficult. Their light output is customarily measured in *millicandelas* (mcd), because LED indicators almost always include a lens that focuses the light, and the candela is a measurement of luminous flux within an angle of dis-

persion. Moreover, because the intensity of neon indicators is not of great interest in most applications, datasheets usually do not supply an intensity value.

One way around the problem of comparisons is to use the standard of *radiant luminous efficacy* (LER), which is defined in the entry on incandescent lamps (see "Efficacy" on page 179). A standard-brightness neon bulb has an LER of about 50 lumens per emitted watt of luminous flux. A light-emitting diode may reach an LER of 100 lm/W. However, a neon bulb operates typically around 1mA while an LED indicator may use 20mA. Therefore, a typical LED indicator may appear to be 30 to 50 times brighter than a typical neon bulb.

Consequently, neon may be an inferior choice in a location where there is a high level of ambient light. Direct sunlight may render the glow of a neon indicator completely invisible.

Efficiency

Because a neon bulb does not use a lot of power and generates negligible heat, it is a good choice where current consumption is a consideration (for example, if an indicator is likely to be on for long periods). The durability and low wattage of neon bulbs, and their convenient compatibility with domestic power-supply voltage, made them a favorite for night-lights and novelty lamps in the past. Figure 19-10 shows an antique bulb containing an ornamental electrode, while Figure 19-11 is a piece of folk art, approximately 1" in diameter, mounted on a plug-in plastic capsule containing a neon bulb.

Ruggedness

Neon bulbs are a good choice in difficult environments, as they are not affected by vibration, sudden mechanical shock, voltage transients, or frequent power cycling. Their operating temperature range is typically from -40 to +150 degrees Celsius, although temperatures above 100 degrees will reduce the life of the lamp.

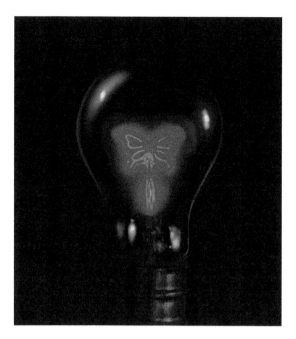

Figure 19-10. *In bygone decades, ornamental neon bulbs with specially shaped electrodes were popular.*

Figure 19-11. *Neon folk art survives in this hand-painted night-light sold in a Florida tourist shop.*

Power-Supply Testing

When driven by DC current, only the negative electrode (the *cathode*) of a neon bulb will glow.

When AC current passes through the bulb, both terminals will glow.

If a bulb (with series resistor) is placed between the "hot" side of a domestic AC power supply and ground, the bulb will glow. If it is placed between the neutral side of the supply and ground, it will not glow.

These features enable a neon bulb to be used for simple power-supply testing.

Life Expectancy

The metal of the electrodes gradually vaporizes during everyday use of a neon bulb. This is known as *sputtering* and can be observed as the glass capsule becomes darkened by deposition of vaporized metal. The electrodes will have a more limited life in a lamp used with DC voltage, where sputtering affects only the cathode. Using AC, the electrodes take turns functioning as the cathode, and vaporization is distributed between both of them.

Failure of a neon lamp can occur as sputtering erodes the electrodes to the point where the maintaining voltage will increase until it almost reaches the level of the power supply. At this point, the bulb will flicker erratically.

Failure can also be defined as a gradual reduction in brightness to 50% of rated light output, caused by accumulated deposition in the glass capsule. Because deposition occurs more heavily on the sides of the bulb, a longer apparent life is possible if the bulb is mounted so that it is viewed from the end.

Typically, neon bulbs are rated for 15,000 to 25,000 hours (two to three years of constant operation). However, the life can be greatly increased by a slight reduction in voltage, which may be achieved by substituting a series resistor with a slightly higher value.

The relationship between operating life and resistor value is shown below. If LA is the normal operating life, LB is the extended operating life,

RA is the normal resistor value, and RB is a higher resistor value:

```
LB = LA * ( RB / RA ) 3.3
```

For example, if a normal resistor value is 20K, and it is increased to 22K, the life of the lamp should increase by a factor of slightly more than 1.4.

Variants

A typical neon bulb terminates in leads, and a lamp assembly often has solder tabs, although it may have a base with a screw thread, flange, or bayonet pins for insertion into a compatible socket. A lamp assembly that does not use a base will either snap-fit into a hole of appropriate size and shape, or may be retained with a nut that engages with a plastic thread on the cylinder of the lamp.

Some neon bulbs or lamp assemblies terminate in pins for direct insertion into a printed circuit board.

Almost all neon bulbs operate either in the 100V to 120V range or in the 220V to 240V range.

Light intensity is expressed either as "standard" or "extra-bright," although datasheets usually do not define those terms.

Nixie Tubes

Nixie tubes, first marketed in 1955, were used to display numerals from 0 through 9 in the days before LEDs took over this capability. They are no longer being manufactured.

Each numeral was physically formed from metal and functioned as an electrode inside a tube filled with a neon-based gas mixture. The typographical elegance of the digits and their aesthetically pleasing glow made Nixies enduringly popular. With a long lifespan, vintage tubes are still usable and can be purchased cheaply from sources such as eBay. Many originate in Russia, where Nixie-type displays were manufactured into the 1980s. The Russian tubes can be identi-fied by their use of a numeral 5 that is a numeral 2 turned upside-down.

Nixie tubes typically require 170VDC. This creates a challenge for a power supply and switching, and can be a safety hazard.

Figure 19-12 shows six Nixie-type tubes repurposed for use as a 24-hour digital clock.

Figure 19-12. *A 24-hour clock using Nixie-type tubes. Source: Wikipedia, public domain.*

What Can Go Wrong

False Indication

Because a neon bulb requires so little power, it may be energized by induced voltages from elsewhere in a circuit, especially if inductive components such as transformers are used. To prevent this, a high-value resistor can be placed in parallel with the bulb, in addition to the series resistor that must always be used.

Failure in a Dark Environment

Because a neon bulb requires a minimal amount of light to initiate its own photon emissions, it may take time to start glowing in a very dim environment, and may not light at all in total darkness. A few bulbs include a small amount of radioactive material that enables them to self-start in complete absence of ambient light.

Premature Failure with DC

The life expectancy quoted in datasheets for neon bulbs usually assumes that they are powered by AC. Because DC results in faster vaporization of the electrodes, the expected lifetime should be reduced by 50% if DC power will be used.

Premature Failure through Voltage Fluctuations

Because the deterioration of a neon bulb accelerates rapidly with current, a sustained voltage that passes slightly more current can radically reduce the expected lifespan.

Replacement

Replacement can be an issue with panel indicators, where disassembly of a device may be necessary to reach the bulb. Bear in mind, however, that an easily removable bulb becomes vulnerable to tampering.

fluorescent light

This entry deals primarily with *fluorescent tubes* (infrequently but sometimes described as *fluorescent lamps*), and *compact fluorescent lamps* (CFLs) that are marketed as a substitute for **incandescent lamps**. *Cold-cathode fluorescent lamps* (CCFLs) are also mentioned.

Vacuum fluorescent devices have a separate entry in this encyclopedia. A fluorescent tube or CFL does not contain a vacuum.

Although the diode(s) in a white **LED area lighting** unit are coated with a layer of fluorescent phosphors, they are not categorized here as fluorescent lights, and have their own entry.

A **neon bulb** resembles a fluorescent light in that it is a *gas-discharge* device, but the interior of its glass envelope is usually not coated with fluorescent phosphors, and therefore it has its own entry.

OTHER RELATED COMPONENTS

- **incandescent lamp** (see Chapter 18)
- **LED area lighting** (see Chapter 23)
- **vacuum fluorescent** devices (see Chapter 25)
- **neon bulb** (see Chapter 19)

What It Does

Fluorescent tubes or *compact fluorescent lamps* (CFLs) are primarily used for area lighting. A partially disassembled CFL appears in Figure 20-1, showing the control electronics that are normally hidden inside the base.

There is no standardized schematic symbol to represent a fluorescent light. Figure 20-2 shows three commonly used symbols for a fluorescent tube on the left, and three symbols for a CFL on the right. Note that two of the symbols for a CFL are the same as those for an **incandescent lamp**, shown in Figure 18-1.

How It Works

Luminescence is the emission of light as a result of a process that does not require heat. (The opposite phenomenon is *incandescence*, in which heating causes an object to emit light; see Chapter 18 for a description of **incandescent lamps**.)

Fluorescence is a form of luminescence. It occurs when electrons in a material are energized and then make a transition back to ground level, at which point they radiate their energy as visible light. The incoming energy can consist of other light at a higher frequency. Some creatures, including species of arachnids and fish, will fluoresce when they are lit with ultraviolet light.

Figure 20-1. *A compact fluorescent lamp with its base cut away to reveal the control electronics.*

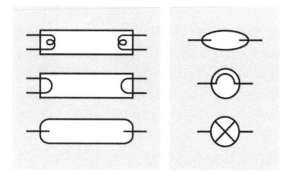

Figure 20-2. *Schematic symbols to represent fluorescent tubes and bulbs are not standardized. See text for details.*

A *fluorescent tube or lamp* contains a very small amount of mercury vapor that can be stimulated to emit ultraviolet light. This encounters a thin layer of *phosphors* coating the inner surface of the glass enclosure. The light causes the phosphors to *fluoresce*, emitting a diffuse radiance in the visible spectrum.

The tube or lamp also contains one or more inert gases such as argon, xenon, neon, or krypton at about 0.3% of normal atmospheric pressure. Two electrodes inside the glass enclosure are made primarily from tungsten, which can be preheated

to initiate ionization of the gas. Confusingly, both electrodes are often referred to as *cathodes*.

The function of the gas is not to emit light, but to conduct electric current, so that free electrons may encounter mercury atoms, raising their electrons briefly to a higher energy level. When one of these electrons reverts from its unstable energized state to its previous energy level, it emits a photon at an ultraviolet wavelength.

Figure 20-3 provides a diagram showing the interior of a fluorescent tube.

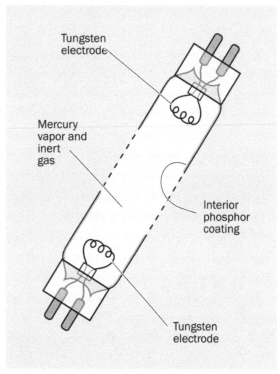

Figure 20-3. *The basic parts of a fluorescent tube.*

Ballast and Starter

Heating the tungsten electrodes is necessary but not sufficient to trigger ionization. A high-voltage pulse is also needed when the light is switched on. In a typical 48" tube, the pulse may range from 200V to 300V.

After current flow has been established, the gas, which is now a plasma, enters a phase of negative resistance. Current passing through it will tend to increase even if the voltage decreases. This process must be controlled to prevent the formation of an *arc*, which will destroy the electrodes. (A similar process occurs in any gas discharge tube, such as a **neon bulb**, and is described in a graph in Figure 19-5.)

To heat the electrodes, ionize the gas, and then control the current, the fixture for a fluorescent tube contains components that are separate from the tube. In their simplest, traditional form, these components consist of a *starter* and a *ballast*. The starter is a neon bulb that contains a bimetallic strip serving as a normally closed switch. It allows current to flow through the electrodes in series, to heat them. The basic circuit is shown in Figure 20-4.

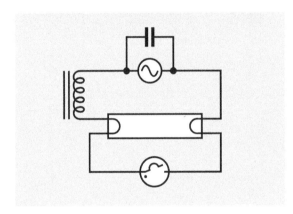

Figure 20-4. *The traditional circuit to trigger ionization of the gas in a fluorescent tube uses a starter (shown at the bottom as a neon tube containing a bimetallic strip, which serves as a switch) and ballast (an inductive load, shown at left).*

The starting process may not be immediately successful, in which case the starter may repeat several times in succession, causing the tube to flicker before its discharge becomes stable. In a cold environment, the tube will have more difficulty starting.

After the tube becomes conductive, current between the electrodes bypasses the starter. At this point, the *ballast* limits the current to prevent an arc from forming. The simplest form of ballast is a coil that functions as an **inductor**.

In a more modern system, an *electronic ballast* replaces the starter-ballast combination. It not only applies the initial surge of high current but also raises the 50Hz or 60Hz frequency of the power supply to 10KHz or more. This increases the efficiency of the tube and eliminates any visible flickering of the light.

All compact fluorescent bulbs (CFLs) contain electronic ballasts. The small components visible in Figure 20-1 are the ballast.

Flicker

When a fluorescent tube uses a conventional ballast and is illuminated with 50Hz or 60Hz AC, the glow discharge stops each time the current flow passes through the zero point in its cycle. In fact, the ionized gas in the tube cannot conduct until it is close to the maximum voltage, and stops conducting when the voltage rolls off. Consequently, the voltage across the tube fluctuates in an approximate square wave, and the light output begins and ends very abruptly. Although this occurs 100 times per second on a 50Hz supply and 120 times per second on a 60Hz supply, some people complain that the flicker is noticeable and can induce headaches.

The rapid on-off discharge is hazardous when it illuminates rotating parts in machinery, as a stroboscopic effect can make the parts seem to be stationary. To mitigate this effect, adjacent tubes in a fixture are powered by separate supplies that are out of phase. This is done either by using a three-phase power supply or by adding an LC circuit to the supply for one of the tubes.

Variants

The traditional type of ballast is also known as a *rapid-start ballast*. By preheating the electrodes, it reduces damage to them that otherwise tends to occur during the starting process. A tube designed for use with a rapid-start ballast has two contacts at each end, and is referred to as a *bi-pin* tube.

An electronic ballast is also known as an *instant-start ballast*. It does not preheat the electrodes, and a tube designed to work with it has only one pin at each end.

CCFLs

A *cold cathode fluorescent lamp* (CCFL) may resemble a miniature fluorescent tube, typically measuring 2mm to 5mm in diameter. The tube may be straight or bent into a variety of shapes. It works on the same principle as a full-size fluorescent tube, containing mercury vapor and one or more inert gases, with an interior coating of phosphors to enable fluorescence. CCFLs are available in many colors and many shades of white.

As its name implies, the electrodes in a CCFL are not heated to establish ionization. Instead, a very high voltage (1,000VAC or more) is applied, dropping to 500VAC to 600VAC after the flow of current has been established. Because CCFLs have been often used to backlight laptop computer screens, inverter circuits are commonly available that create a high-frequency output at a high voltage from an input that can range from 3VDC to 20VDC. The inverter also includes provision to dim the CCFL by using *pulse-width modulation*.

Some CCFLs are designed for illumination of small spaces—for example, the interior of a display case. A few CCFLs look exactly like CFLs and can be used in light fixtures. Some may be compatible with the type of dimmer designed for incandescent lamps.

A CCFL usually has a limited light output compared with that of a conventional fluorescent tube, but has the advantage of working better at low temperatures. Some are designed for signage and exterior lighting in cold-weather locations.

They have a relatively long lifetime of up to 60,000 hours. A hot-cathode fluorescent lamp may fail between 3,000 and 15,000 hours.

Any tube or bulb that uses unheated electrodes to ionize a gas is technically a cold-cathode device, but will not be identified as a CCFL unless it also has an inner layer of phosphors to achieve fluorescence.

It is important to match a tube with the type of ballast installed in a fixture. This is not an issue with CFLs, as they have the appropriate ballast built in.

Sizes

Straight bi-pin tubes are sold in the United States in the following standard sizes:

- T5: 5/8" diameter. A more modern tube, but still with tungsten electrodes that serve to heat it.
- T8: 1" diameter. Very often 24" or 48" in length, consuming 18W or 36W respectively.
- T12: 1-1/2" diameter.
- T17: 2-1/8" diameter.

CFLs are sold in a very wide variety of configurations.

Comparisons

Fluorescent lights have significant advantages and disadvantages. On the plus side:

- After the fixture containing the ballast has been paid for, a tube is relatively cheap. A CFL or an LED light does not have this advantage, as the electronics are built in and will be discarded when the light fails.
- Fluorescent lights have a longer life than incandescent bulbs.
- Fluorescent lights are available in a wide range of shades of white.
- Fluorescent tubes create a diffuse radiance that is ideal for general lighting using ceiling-mounted fixtures. They do not cast harsh shadows.

On the minus side:

- Fluorescents were traditionally more energy-efficient than any other light source, but **LED area lighting** is now more efficient in some designs. LEDs are expected to become more efficient in the future.

- A fluorescent tube with a traditional type of ballast may cause complaints of flickering. By comparison, an LED light uses DC, and an incandescent bulb retains sufficient heat between power cycles so that it does not appear to flicker.

- Fluorescent flicker creates problems when shooting video.

- The fluorescent emission spectrum has sharp peaks that give the lighting an unnatural look.

- In applications that require a defined beam of light, a fluorescent source cannot be used.

- Conventional ballasts can create radio interference, especially in the AM band.

- Because fluorescent tubes and bulbs contain mercury, they require proper disposal, which can incur fees.

- Even an instant-on fluorescent light tends to hesitate briefly when it is switched on.

- The lifespan of a fluorescent light is greatly reduced if it is cycled on and off frequently. An incandescent bulb is less severely affected by cycling, and an LED light is not affected at all.

- Fluorescent lights have difficulty starting at low temperatures.

Values

Brightness

The intensity of a fluorescent light is measured in *lumens per watt*. Because invisible wavelengths are of little interest when assessing brightness, *luminous flux* is used to describe apparent brightness in the visible spectrum. The unit for luminous flux is the *lumen*. Additional information about light measurement is included in the entry describing incandescent lamps (see "Power" on page 177).

Spectrum

The spectrum of photons emitted from mercury vapor in a fluorescent light has wavelengths that peak at 253.7 nanometers and 185 nanometers. (A *nanometer*, customarily abbreviated as nm, is one-billionth of a meter.) These wavelengths are invisible, being in the ultraviolet range, but when the light is transposed into the visible spectrum by the layer of phosphors, "spikes" in the range of wavelengths are still present. For a comparison of output curves for incandescent, fluorescent, and LED lights, see the graph in Figure 18-4.

Various formulations for the phosphors in a tube or CFL attempt to modify the character of the light to suit the human eye, but none of them looks as "natural" as the radiance from an incandescent bulb, probably because the characteristics of incandescent light are very similar to those of sunlight.

What Can Go Wrong

Unreliable Starting

At a low temperature, the mercury inside a fluorescent tube may be slow to vaporize. At very low temperatures, vaporization may not be possible at all. Until the mercury vaporizes, fluorescence will not occur.

Terminal Flicker

As a tube ages, it may start to conduct current only in one direction, causing it to flicker visibly. As it ages more, the gas discharge becomes even less reliable, and the flicker becomes erratic. Eventually, the gas discharge fails completely. In this state, a tube may show only a dim light at each end, in proximity to the tungsten electrodes.

Cannot Dim

Neither the older style of "conventional" ballast nor a modern electronic ballast will respond appropriately to a dimmer of the type designed for incandescent bulbs. This may be an important factor when an incandescent bulb is swapped out for a CFL.

Burned Out Electrodes

Like the tungsten filament in an **incandescent lamp**, the tungsten electrodes in a fluorescent tube suffer progressive erosion. This is evident when a black tungsten deposit forms on the inside of the tube at one or both ends.

Ultraviolet Hazard

Some critics of CFLs maintain that the complex shape of a coiled or zig-zag tube tends to permit small imperfections in the internal phosphor coating, potentially allowing ultraviolet light to escape. If this occurs, and if a CFL is used in a desk fixture in close proximity to the user, ultraviolet light could elevate the risk of skin cancer.

laser | 21

The term *maser* was coined in the 1950s to describe a device that used stimulated emission to amplify microwaves. When a device using similar principles amplified visible light in 1960, it was termed an *optical maser*. However, that term is now obsolete, having been replaced with **laser**. This term is always printed in lowercase letters, even though it is an acronym for Light Amplification by Stimulated Emission of Radiation.

The invented verb *to lase* is derived from *laser* and is used to describe the process of generating laser light, with the past participle *lased* and present participle *lasing* sometimes being used.

Thousands of laser variants exist. Because of space limitations, this entry will concentrate primarily on *laser diodes*, which are the smallest, most common, and most affordable type.

OTHER RELATED COMPONENTS

• **LED indicator** (see Chapter 22)

What It Does

A **laser** generally emits a thin beam of intense light, often in the visible spectrum, and usually in such a narrow range of wavelengths, it can be considered *monochromatic*. The light is also *coherent*, as explained below.

Light output from a laser has three important attributes:

• Intensity. A high-powered laser can deliver energy to a very small, well-defined area, where it may be capable of burning, cutting, welding, or drilling. Large lasers may also be used as weapons, or for power transmission.

• Collimation. This term describes a beam of light that has parallel boundaries, and therefore does not disperse significantly when passing through a transparent medium such as air, glass, or a vacuum. A laser beam can have such excellent collimation, it can be used in precision measuring devices, and has been transmitted over very long distances, even from the Earth to the Moon, where astronauts placed reflectors during the Apollo missions.

• Controllability. Because the beam can be generated with eletrical power, its intensity can be modulated rapidly with relatively simple electronic circuits, enabling applications such as burning microscopic pits in the plastic of a CD-ROM or DVD.

Laser diodes are now more common than all other forms of lasers. They are found in pointers, printers, barcode readers, scanners, computer mice, fiber-optic communications, surveying tools, weapon sights, and directional lighting sources. They are also used as a light source to trigger more powerful lasers.

No generic symbol is used for a laser, but a laser diode is often represented with the same symbol that is used for a light-emitting diode. See Figure 22-2 in the entry for **LED indicators**.

How It Works

A laser is built around a *gain medium*, which is a material that can amplify light. The medium can be a solid, liquid, gas, or plasma, depending on the type of laser.

Initially, an input of energy provides stimulation for some atoms in the gain medium. This is known as *pumping* the laser. The energy input can come from a powerful external light source, or from an electric current.

Stimulation of an atom raises the quantum energy level of an electron associated with the atom. When the electron collapses back to its former energy state, it releases a photon. This is known as *spontaneous emission*.

If one of the photons encounters an atom that has just been excited by the external energy source, the atom may release two photons. This is known as *stimulated emission*. Beyond a threshold level, the number of released photons can increase at an exponential rate.

If two parallel reflectors are mounted at opposite ends of the gain medium, they form a *resonant cavity*. Light bounces to and fro between the reflectors, while pumping and stimulated emission amplifies the light during each pass. If one of the mirrors is partially transparent, some of the light will escape through it in the form of a laser beam. The partially transparent mirror is known as the *output coupler*.

Laser Diode

A laser diode contains an **LED**. (See "How It Works" on page 207 for a more detailed description of the function of an LED.) The p-n junction of the diode functions as the resonant cavity of the laser. Forward bias injects charges into the junction, causing spontaneous emission of photons. The photons, in turn, cause other electrons and electron-holes to combine, creating more pho-

tons in the process of stimulated emission. When this process crosses a threshold level, current passing through the diode causes it to *lase*.

The original patent for a laser diode was filed by Robert N. Hall of General Electric in 1962, and the diagram in Figure 21-1 is derived from the drawing in that patent, with color added for clarity.

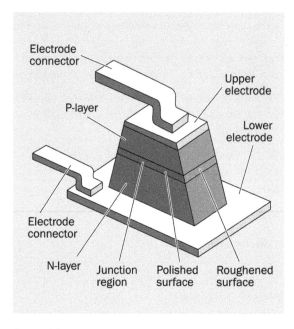

Figure 21-1. *The original design for a laser diode, from the patent filed in 1962.*

In the figure, the junction shown as a yellow layer forms the resonant cavity in which lasing occurs. It measures only 0.1 microns thick (the diagram is not drawn to scale). Its vertical front side is highly polished, and is parallel to the back side, which is also highly polished. Thus, photons reflect between these two vertical sides. The slanted face visible in the figure, and the other slanted face opposite it, are oriented and roughened to minimize internal reflection between them.

Figure 21-2 shows a simplified cross-section of the laser diode.

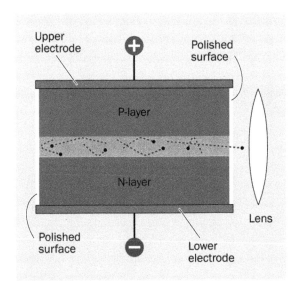

Figure 21-2. *Simplified cross-section of a laser diode.*

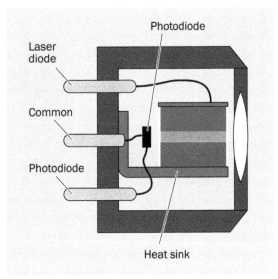

Figure 21-3. *A laser diode is typically mounted with a photodiode to provide feedback for a driver circuit, to control the current consumed by the laser.*

Figure 21-3 shows a cross-section of the diode installed in a component sold as a laser. It includes a photodiode to sense the intensity of light emerging through the polished rear end of the laser diode. External electronics are necessary to control the intensity of the laser, using feedback from the photodiode.

The component has three pins (shown pale yellow in the figure), one connecting to the photodiode, another connecting to the p-type layer of the laser diode, and the third being common to the n-type layer of the laser diode and the ground side of the photodiode.

A photograph of a laser diode is shown in Figure 21-4. Note the three pins, comparable to the pins shown in Figure 21-3, indicating that this component requires external control electronics.

In Figure 21-5, a laser is shown with a surface-mount chip adjacent to the solder pad connecting the blue wire. The presence of this chip, with only two wires, indicates that this component has its own control electronics and requires only a DC power supply.

Figure 21-4. *Lite-On 505T laser diode that emits light at 650nm. Power consumption 5mW at 2.6VDC. As indicated by the graph squares, this component is only about 0.2" in diameter.*

suggested in Figure 21-6 where the light source is an incandescent lamp emitting a wide range of wavelengths.

Figure 21-5. *This laser incorporates its own control electronics and requires only a 5VDC power supply. It draws 30mA and generates an output up to 5mW.*

Coherent Light

The emission of *coherent light* by a laser is often explained by suggesting that wavelengths are synchronized with each other. In fact, there are two forms of coherence that can be described approximately as *spatial coherence* and *wavelength coherence*.

If an observer looks up at a cloudy sky, the eye will perceive light radiating chaotically from many distances and directions. Thus, the light is not spatially coherent. The light also consists of many wavelengths, and thus it is not wavelength-coherent.

The filament of an incandescent lamp is a much smaller source of light, but still large enough to generate a profusion of light emissions that are spatially incoherent. The light also includes many different wavelengths.

Suppose a barrier containing a very tiny hole is placed in front of the **incandescent lamp**. If the aperture is very small, an observer on the far side will see the light as a point source. Consequently, the light that emerges from it is now spatially coherent, and will not have chaotically overlapping waves. If the light then passes through a filter, its wavelengths also will become coherent. This is

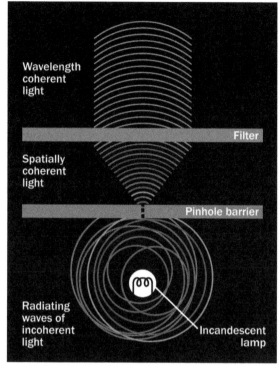

Figure 21-6. *An incandescent lamp, at the bottom of the figure, emits incoherent light at many wavelengths (exaggerated here for clarity). When it passes through a pinhole, it becomes spatially coherent. When it then passes through a colored filter, it becomes wavelength coherent.*

The small amount of light emerging through a pinhole is inevitably much dimmer than light from the original source. A laser, however, amplifies its light output, as well as tending to behave like a point source. The "hall of mirrors" effect of the parallel reflective surfaces in the resonant cavity causes much of the light to shuttle to and fro over a long distance before it emerges through the output coupler. Any light that deviates significantly from the axis of the laser will not escape at all, because the deviations will be cumulative with each reflection. Thus, the light from a laser appears to come from a point source at an almost infinite distance.

Because of the particular geometry of a light-emitting diode, the output from a laser diode is not naturally collimated, and tends to spread by an angle of around 20 degrees. A lens must be used to focus the beam.

Variants

Lasers are sold generally as fully assembled tools for a specific purpose. A very brief summary of CO2, fiber, and crystal lasers is included here.

CO2 Lasers

The gain medium is primarily carbon dioxide but also contains helium and nitrogen, with sometimes hydrogen, water vapor, and/or xenon. The laser is electrically pumped, causing a gas discharge. Nitrogen molecules are excited by the discharge and transfer their energy to the CO2 molecules when colliding with them. Helium helps to return the nitrogen to base energy state and transfer heat from the gas mixture.

CO2 lasers are infrared, and are commonly used in surgical procedures, including ophthalmology. Higher powered versions have industrial applications in cutting a very wide range of materials.

Fiber Lasers

Light is pumped via diodes and amplified in purpose-built glass fibers. The resulting beam has a very small diameter, providing a greater intensity than CO2 lasers. It can be used for metal engraving and annealing, and also for working with plastics.

Crystal Lasers

Like fiber lasers, they are pumped by diodes. These compact lasers are available in a very wide variety of wavelengths, covering the whole visible spectrum, infrared, and ultraviolet. They find applications in holography, biomedicine, interferometry, semiconductor inspection, and material processing.

Values

The output power of a laser is measured in watts (or milliwatts). This should not be confused with the power consumed by the device.

In the United States, any device sold as a laser pointer is limited to a power output of 5mW. However, laser diodes packaged similarly to laser pointers can be mail-ordered with an output of 200mW or more. The legal status of these lasers may be affected by regulations that vary state by state.

In a CD-RW drive that is capable of burning a disc, the diode may have a power of around 30mW. A laser mounted in a CD-ROM assembly is shown in Figure 21-7.

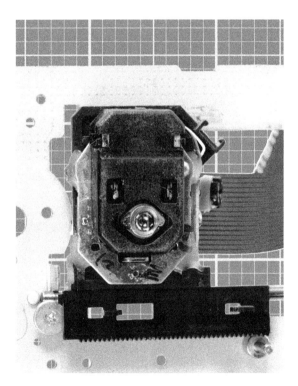

Figure 21-7. *An assembly incorporating a laser for reading a CD-ROM.*

Lasers have such a narrow range of wavelengths, they are given specific output values in nanometers. A laser in an optical mouse may have a wavelength of 848nm; in a CD drive, 785nm; in a bar-code reader, 670nm; in a modern laser pointer, 640nm; in a Blu-ray disc player, 405nm.

How to Use It

While powerful lasers in a laboratory setting have exotic applications, a typical low-power laser diode has become so affordable (costing less than $5 in some instances, at the time of writing) it can be considered merely as a useful source of a clearly defined light beam, ideal for detecting the position of a movable mechanical component or the presence of an intruder.

Generic light-emitting diodes are made with a view angle (i.e., a dispersion angle) as low as 3 degrees, but the beam is soft-edged compared with the precise boundary of a laser beam, and cannot be used reliably in conjunction with sensors at a distance of more than a few inches.

Laser diodes that are sold as components may or may not have current-limiting control electronics built in. Applying power to the laser diode directly will result in thermal runaway and rapid destruction of the component. Drivers for laser diodes are available separately as small, preassembled circuits on breakout boards.

For many applications, it may be simpler and cheaper to buy a laser diode as an off-the-shelf product. A laser pointer provides an easy way to get a source of laser light, and if it would normally be driven by two 1.5V batteries, it can be adapted to run off a 5V supply by using a 3.3V voltage regulator.

Common Applications

In addition to being used with PowerPoint presentations and in conjunction with position sensors, laser pointers have other applications:

- Astronomy. A high-powered laser beam is visible even in clear air as a result of interaction with air molecules. This is known as *Rayleigh scattering*. The phenomenon allows one person to point out a star (or planet) for another person. Because celestial objects are so far away, parallax error is not detectable by two people viewing the beam while standing next to each other. A laser pointer may also be mounted on a telescope to assist in aiming the telescope at an object of interest. This is easier than searching for an object through an eyepiece.

- Target acquisition. Lasers are commonly used on firearms to assist in targeting, especially in low-light conditions. Infrared lasers can be used in conjunction with infrared viewing goggles.

- Survival. A small laser can be included in emergency supplies to signal search teams. A laser can also be used to repel predatory animals.

What Can Go Wrong

Risk of Injury

Lasers are potentially dangerous. Those that have an infrared or ultraviolet output are more dangerous than those with a visible beam, as there is no visual warning that the laser is active. A laser is capable of scarring the retina, although controversy exists regarding the power output that should be considered a high risk.

If a project incorporates a laser, it should be switched off while building or testing the device. It may be advisable to wear protective glasses that block laser light even when an experimenter feels confident that a laser is switched off.

Active lasers should never be pointed at people, vehicles, animals (other than dangerous animals), or oneself.

Inadequate Heat Sink

Lasers may be designed and rated for intermittent use. The burner assembly for a CD-ROM drive, for instance, will be rated for pulsed power, not continuous power. Read datasheets carefully, and provide an adequate heat sink.

Uncontrolled Power Supply

A diode laser that does not have a feedback system in place to control the flow of current can self-destruct.

Polarity

Both the light-emitting diode and the photodiode in a three-pin laser package can be damaged by incorrect polarity of applied power. Pin functions should be checked carefully against datasheets.

LED indicator

22

In this encyclopedia, an **LED indicator** is defined as a component usually 10mm or smaller in diameter, made of transparent or translucent epoxy or silicone, most often containing one *light-emitting diode*. It is purposed as a status indicator in a device, rather than as a source of illumination, and is sometimes referred to as a *standard LED.*

LED indicators that emit infrared and ultraviolet light are included in this entry. LEDs that are designed to illuminate large interior or exterior areas are discussed in a separate entry as **LED area lighting**. They are sometimes described as *high-brightness LEDs* and almost always emit white light.

The term **light-emitting diode** is becoming less common, as the acronym **LED** has become ubiquitous. The acronym does not usually include periods between the letters.

The words "light emitting" are hyphenated here, as they form an adjectival phrase, but in everyday usage the hyphen is often omitted, and no definitive rule seems to exist.

Originally, a standard LED contained only one diode, but may now include multiple diodes, either to emit additional light or to provide a range of colors. In this encyclopedia, a single epoxy or silicone capsule is still considered to be an **LED indicator** regardless of how many diodes it contains. By contrast, any component consisting of multiple separately discernible light-emitting diodes, as in a seven-segment numeral, a 14- or 16-segment alphanumeric character, a dot-matrix character, or a display of multiple characters, is listed in a separate entry as an **LED display**.

OTHER RELATED COMPONENTS

- **LED area lighting** (see Chapter 23)
- **LED display** (see Chapter 24)
- **incandescent lamp** (see Chapter 18)
- **neon bulb** (see Chapter 19)
- **laser** (see Chapter 21)

What It Does

An **LED indicator** emits light in response to a small current, typically around 20mA (but sometimes much less), at a voltage lower than 5VDC. It is usually molded from epoxy or silicone that may be colorless and transparent (often referred to as *water clear*), or colorless but translucent, or tinted and transparent, or tinted and translucent.

The color of the light is initially determined by the chemical compounds used internally, and by their dopants; therefore, a water-clear LED may emit colored light.

Ultraviolet LEDs are usually water-clear. Infrared LEDs often appear to be black, because they are opaque to the visible spectrum while being transparent to infrared.

When an LED indicator is described as being *through hole,* it has leads for insertion into holes in a circuit board. The term does not mean that the indicator itself is meant to be pushed through a hole in a panel, although this may also be done. The LED is cylindrical with a hemispherical top that acts as a lens. The leads are relatively thick, to conduct heat away from the component. A dimensioned diagram of a typical LED measuring 5mm in diameter is shown in Figure 22-1.

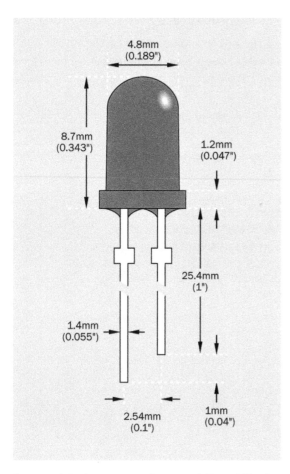

Figure 22-1. *Dimensions of a typical 5mm LED. The longer wire connects with the anode, while the shorter wire connects with the cathode. Adapted from a datasheet published by Lite-On Technology Corporation.*

An LED indicator that is not the through-hole type is usually a surface-mount component. LEDs for surface mounting are mostly rectangu-

lar and can be as small as 1mm x 0.5mm. They may require a *heat sink.*

Schematic Symbols

Figure 22-2 shows a variety of symbols that are commonly used to represent an LED. The triangle at the center of each symbol points in the direction of conventional (positive-to-negative) current flow—from the anode to the cathode. Each pair of arrows radiating away from the diode indicates emitted light. Wavy arrows are sometimes used to represent infrared (thermal) radiation. Often, however, an infrared LED is represented in exactly the same style as an LED that emits visible light. With the exception of the wavy arrows, the various styles of schematic symbol are functionally identical and do not identify different attributes of the component such as size or color.

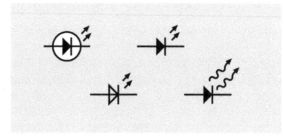

Figure 22-2. *Various symbols may be used to represent an LED. See text for details.*

Common Usage

LED indicators have mostly displaced **neon bulbs** and miniature **incandescent lamps** for the purpose of showing the status of a device. They are found in industrial control panels, home audio systems, battery chargers, washer/dryers, and many other consumer-electronics products. Higher output variants are used in flashlights, traffic signals, taillights on automobiles, and for illumination of subjects that are being photographed. LED indicators may be assembled in large numbers in attention-getting displays such as Christmas lights.

Red, orange, yellow, green, and blue are the basic standard colors. LEDs that appear to generate white light are common, but they do not emit an evenly weighted spectrum of wavelengths. See "Non-Incandescent Sources" on page 174 for a discussion of this topic.

How It Works

An LED, like any **diode**, contains a semiconductor *PN junction* that conducts current only in the forward direction (i.e., from the more-positive side of a power supply to the more-negative side). The diode becomes conductive above a *threshold voltage* sufficient to force electrons in the n-type region and holes in the p-type region to combine with each other. Each time this occurs, energy is released. The energy liberated by one electron-hole combination creates a *photon*, or one quantum of light.

The amount of energy released depends on the *band gap*, which is a property of the semiconductor material. The band gap is the smallest energy that can create an electron-hole pair. The energy determines the light's wavelength, and thus the color.

The band gap also determines the threshold voltage of the LED. For this reason, LEDs of different colors have widely different threshold voltages.

Because an LED will often be used in devices where the DC power supply exceeds the maximum forward voltage, a *series resistor* is customarily used as a simple way to restrict current through the diode.

The light emitted by a colored LED indicator tends to include only a narrow range of wavelengths. However, the addition of a phosphor coating to the diode can broaden the output. This technique is used to make the light from a blue LED appear white, as shown in Figure 22-3. Most white LEDs are actually blue LEDs with a colored phosphor layer added. See the section

on **LED area lighting** in Chapter 23 for a more detailed discussion of this topic.

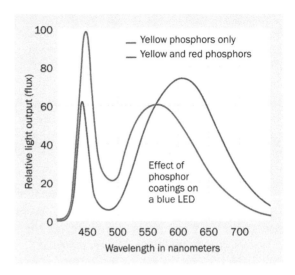

Figure 22-3. *Increasing the range of emitted wavelengths by adding phosphors to a blue LED. Source: Philips Gardco Lighting.*

Multicolor LEDs and Color Mixing

If red, green, and blue light sources are mounted extremely close together, the eye perceives them as a single source, of a color determined by their combined relative intensities. This system of additive color mixing is shown in Figure 17-14 in the entry dealing with LCDs. It is used in LED indicators that contain red, green, and blue light-emitting diodes in a single epoxy or silicone capsule.

While most video monitors use white LEDs or **fluorescent lights** to form a backlight for an **LCD** video screen, some high-end monitors use a matrix of very tiny red, green, and blue LEDs, because the combination of these separate colors generates a wider *gamut* of color wavelengths. The concept of gamut is discussed in "Color" in the **LCD** entry. The tiny LEDs in a backlight cannot be considered as indicators, but indicators are used for this purpose in billboard-sized video displays.

Variants

LED indicators vary widely in size, shape, intensity, view angle, diffusion of light, wavelength of light, minimum and maximum forward voltage, and minimum and maximum forward current.

Size and Shape

The original sizes for round LED indicators were 3mm, 5mm, or (more rarely) 10mm in diameter. Today, through-hole LEDs are sold in many intermediate sizes, although 3mm and 5mm are still most widely used.

The traditional round LED indicator is now augmented with square and rectangular shapes. In a parts catalog, a pair of dimensions such as 1mm × 5mm suggests that the LED is rectangular.

Intensity

The light intensity of an LED is usually expressed in *millicandelas*, abbreviated *mcd*. There are 1,000 mcd in a candela. For more information about units for measurement of light, see "Intensity" on page 178.

The candela measures the *luminous flux*, or visible radiant power, contained within a specified angle of dispersion, usually referred to as the *view angle*. This can be imagined as the rotated angle at the apex of a cone, where the cone defines the "spread" of the light, and the source is at the apex.

If a diode is emitting a fixed amount of luminous flux, the rating in mcd will increase with the inverse square of the view angle. This is because the light delivered to an area in front of the LED will become more intense as the angle becomes smaller. The use of mcd to rate the brightness of an LED can be misleading if it is not considered in comparison with the view angle.

For example, suppose an LED is rated at 1,000 mcd and has a view angle of 20 degrees. Now suppose the same diode is embedded in a different epoxy or silicone capsule with a lens that creates a view angle of only 10 degrees. The LED

will now be rated at 4,000 mcd, even though its total power output is unchanged.

- To compare the brightness of two LED indicators meaningfully, they should share the same view angle.

Four through-hole LED indicators with a wide range of specifications are shown in Figure 22-4. From left to right: water-clear white generic, 10mm; Vishay TLCR5800 5mm (emitting red, even though the capsule is water-clear), rated for 35,000mcd with 4 degrees view angle; Everlight HLMPK150 5mm red diffused, rated for 2mcd with 60 degrees view angle; and Chicago 4302F5-5V 3mm green, rated for 8mcd at 60 degrees view angle, containing its own series resistor to allow direct connection with a 5VDC power supply.

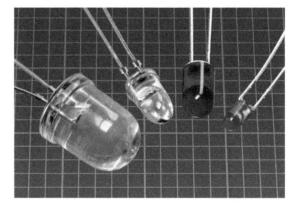

Figure 22-4. *Four assorted LED indicators with very different specifications. See text for details.*

Efficacy

The *radiant luminous efficacy* (LER) of an incandescent light source compares how effective it is at channeling its output within the visible spectrum, instead of wasting it in other wavelengths, especially infrared. Note that the word *efficacy* has a different meaning from *efficiency*. The LER acronym may help to avoid confusion.

LER is expressed in lumens per watt, and in an incandescent bulb it is calculated by dividing the power emitted in the visible spectrum (the *lumi-*

nous flux) by the power emitted over all wavelengths. This is described in detail on "Efficacy" on page 179 in the entry dealing with incandescent lamps.

In an LED indicator, almost all the radiation can be within the visible spectrum, but some power is wasted by generating heat internally. The efficacy varies depending on the type of LED; thus a red-orange indicator can have an efficacy of 98% while a blue LED will be probably below 40%.

Diffusion

Some LED indicators use epoxy or silicone that is formulated to be translucent or "cloudy" instead of transparent. They diffuse the light so that it is not projected in a defined beam, has a softer look, and has an approximately equal intensity when viewed from a wider range of angles.

"Clear" and "diffused" are options that must be taken into account when choosing LEDs from an online catalog, unless the user is willing to turn a clear LED into a diffuse LED by applying some sandpaper.

Wavelength and Color Temperature

The wavelength of light is measured in *nanometers* (abbreviated *nm*), a nanometer being 1 billionth of a meter. The visible spectrum extends from approximately 380nm to 740nm. Longer wavelengths are at the red end of the spectrum, while shorter wavelengths are at the blue end.

A typical LED emits a very narrow range of wavelengths. For example, Figure 22-5 shows the emission from a standard red LED indicator manufactured by Lite-On. Graphs of this type are typically included in manufacturers' datasheets.

Because a red LED stimulates the cones in the eye that respond to red light, it "looks red" even though the color is not comparable with the natural red that is seen, for instance, in a sunset. That natural color actually contains an additional spread of wavelengths.

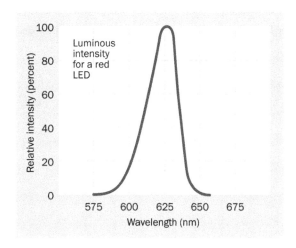

Figure 22-5. *The narrow range of wavelengths emitted by a typical red LED indicator.*

The following list shows the ranges of peak output values, in nanometers, for the most commonly available basic LED indicators (LEDs that emit other wavelengths are available, but they are less common):

- Infrared LED: 850 to 950
- Red LED: 621 to 700
- Orange LED: 605 to 620
- Amber LED: 590 to 591
- Yellow LED: 585 to 590
- Green LED: 527 to 570
- Blue LED: 470 to 475
- Ultraviolet LED: 385 to 405

Figure 22-6 shows this list graphically, omitting infrared and ultraviolet LEDs.

For almost 30 years, blue LEDs were a laboratory curiosity of little practical value, as efficiencies were stuck around 0.03%. An efficiency of more than 10% was finally achieved in 1995. Blue LEDs were marketed soon afterward.

However, when yellow phosphors are added to create the impression of white light by spreading the output over the whole visible spectrum, the wavelengths around 500nm are still not well represented, as suggested in Figure 22-3.

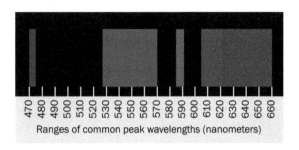

Figure 22-6. *Ranges for peak wavelengths of the most commonly used LEDs. (Source: Survey of approximately 6,000 through-hole LEDs stocked at www.mouser.com.)*

Fluorescent lights perform even more poorly than white LEDs, as can be seen in Figure 18-4 in the entry describing **incandescent lamps**.

Because white LEDs do not emit a single peak of wavelengths, their color is expressed in color temperature rather than nanometers. The concept of color temperature is explained in "Spectrum" on page 173. White LEDs are available rated from 2,800 to 9,000 degrees Kelvin, and are discussed in more detail in the **LED area lighting** entry in this encyclopedia.

Internal Resistor

To eliminate the chore of adding a series resistor to limit current through an LED, some indicators are sold with a series resistor built in. They may be rated for use with 5VDC or 12VDC, but are externally indistinguishable from each other. They are also externally indistinguishable from LEDs that do not contain series resistors. Figure 22-7 shows two 3mm LEDs, the one on the right containing its own series resistor, the one on the left being a generic LED without a series resistor.

Because of the nonlinear response of a diode, LEDs with or without internal resistors cannot be distinguished from each other reliably with a multimeter. If the meter is set to measure ohms, typically it will give an "out of range" error to all types of LED. If it is set to identify a diode, the reading will not tell you if the LED contains a resistor.

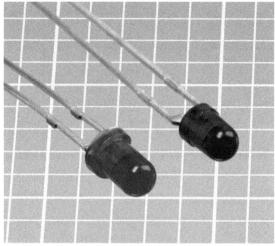

Figure 22-7. *An LED (left) that does not contain its own series resistor is usually indistinguishable from one that does (right).*

One way to determine whether an LED contains an internal series resistor would be to connect it with a variable power supply through a multimeter set to measure mA. Carefully increase the supply voltage from zero until the meter shows a current of 20mA. If the LED does not contain a series resistor, the supply voltage will be close to the recommended forward voltage for that type of LED (no lower than 1.6V for a red LED, and no higher than 3.6V for a white LED). If the LED does contain a series resistor, the supply voltage will be higher. This procedure is time consuming, but may be worthwhile to evaluate multiple LEDs that are known to be identical.

Multicolored

The leads for an LED indicator containing two or more diodes can be configured in several ways:

- Two leads, two colors. Two diodes are mounted internally in parallel, but with opposite polarity.

- Three leads, two colors. Two diodes share a common anode or common cathode.

- Four leads, three colors (RGB). Three diodes share a common anode or common cathode.

- Six leads, three colors. Three diodes, each with its own pair of leads, separate from the others.

Infrared

Most infrared emitters are LEDs that generate wavelengths longer than 800nm. They are found in handheld remotes to control consumer-electronics devices such as televisions and stereo systems, and are also used in some security systems, although *passive infrared motion detectors*, which assess infrared radiation from sources such as people or vehicles, are more commonly used for this purpose.

In conjunction with an infrared emitter, an infrared sensor is necessary, and must be sensitive to the same wavelength. To prevent false positives, the emitter modulates its output, typically with a carrier frequency between 10 and 100kHz. Remotes often use carrier frequencies of 30 to 56kHz. At the receiving end, the signal is processed with a band-pass filter matching the modulation frequency. Many different pulse-coding schemes are used, and no particular standard is dominant.

Ultraviolet

Because ultraviolet radiation can damage the eyes, LED indicators that emit ultraviolet light are potentially dangerous and should be used with caution. A yellow eyeshield can be worn to block the short wavelengths.

Ultraviolet light can be used to cure some adhesives and dental filling material. It can also kill bacteria, and can detect fluorescent print on bank notes, to check for counterfeiting. Ultraviolet flashlights are sold to detect some species of pests, such as scorpions, which fluoresce in response to ultraviolet light.

Values

The specification for an LED will include the wavelength of emitted light, luminous intensity, maximum forward voltage and current, maxi-

mum reverse voltage and current, and working values for voltage and current. All these values are important when choosing an indicator for a specific function.

White LEDs for room lighting or external use are calibrated differently. See the entry for **LED area lighting** in Chapter 23.

Forward Current

About half of all the thousands of available types of LED indicators are rated for a typical forward current of 20mA to 25mA. Absolute maximum ratings may be twice as high, but should not generally be applied.

The light intensity of a typical 5mm red LED indicator is plotted against its forward current in Figure 22-8. Note that current and light intensity have an approximately linear relationship up to the typical working current of 20mA. Even above this point, to the absolute maximum of 50mA, the light intensity rolls off only a very small amount.

Although an LED indicator can be dimmed by controlling the current passing through it, the current does not have a linear relationship with the applied voltage, and the indicator will stop functioning completely when the voltage drops below the threshold required by the diode. Consequently, LEDs are commonly dimmed by using *pulse-width modulation*.

Because of the nonlinear response of a diode, LEDs with or without internal resistors cannot be distinguished from each other reliably with a multimeter. If the meter is set to measure ohms, typically it will give an "out of range" error to all types of LED. If it is set to identify a diode, the reading will not tell you if the LED contains a resistor.

Low-Current LEDs

Indicators that require a very low forward current are convenient for direct connection to output pins of logic chips and other integrated circuits. Although a single output from an HC family chip is capable of supplying 20mA without damaging the chip, the current will pull down the output voltage, so that it cannot be used reliably as an input to another chip while also lighting the LED.

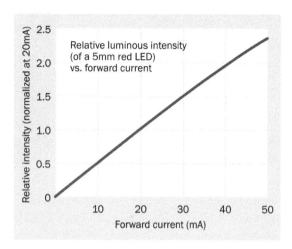

Figure 22-8. *The relationship between forward current and light intensity of a typical 5mm LED indicator is approximately linear up to the 20mA operating current, and almost linear up to the absolute maximum of 50mA.*

Various LED indicators drawing 2mA or 1mA are available, with intensities typically ranging between 1.5mcd and 2.5mcd. This low light output is still bright enough for viewing in a lab-bench environment. Low-current blue LEDs are not available. The only LEDs that draw as little as 1mA are red, as this is the most efficient type.

Using a higher value series resistor with a generic LED will of course reduce its current consumption, and some light will be visible so long as the forward voltage across the LED remains at its minimum level or above.

Forward Voltage

Red is the color that requires not only the least forward current, but the lowest forward voltage. In the range of 1.6VDC to 1.7VDC, all the LEDs are red. Typical forward voltages for various colors are shown here:

- Infrared LED: 1.6V to 2V
- Red LED: 1.6V to 2.1V

- Orange LED: 1.9V to 2.1V
- Amber LED: 2V to 2.1V
- Yellow LED: 2V to 2.4V
- Green LED: 2.4V to 3.4V
- Blue LED: 3.2V to 3.4V
- Ultraviolet LED: 3.3V to 3.7V
- White LED: 3.2V to 3.6V

Color Rendering Index

The *color rendering index* (CRI) evaluates the fidelity with which a light source is capable of displaying the full visible spectrum. It ranges from a perfect score of 100 down to 0 or even lower (sodium-vapor street lighting has a negative value). Computing the index requires standard reference color samples and has been criticized for generating scores that do not correlate well with subjective assessments.

Incandescent bulbs can have a CRI of 100, while an uncorrected white LED may score as low as 80.

Life Expectancy

Because the light output from an LED tends to decrease very gradually with time, the life expectancy is often defined as the number of hours required for the output to diminish to 70% of its output when new. Life expectancy is commonly stated on datasheets for high-brightness white LEDs, but is often omitted from datasheets for LED indicators.

Unlike incandescent lamps and fluorescent lights, LEDs do not have a shorter lifespan if they are frequently cycled on and off.

Light Output and Heat

The light intensity of an LED, measured in mcd, can vary from a few mcd to a maximum of 40,000mcd. Intensities above 30,000mcd generally are achieved by limiting the view angle to 15 degrees or less. Because the candela is weighted toward the central, green segment of the visible spectrum, green LEDs are likely to have a relatively high mcd rating. LEDs rated between 20,000mcd and 30,000mcd, with a view angle of 30 degrees, are almost all green.

Datasheets may often include a *derating curve* showing the lower limit that should be placed on forward current through an LED indicator when its temperature increases. In Figure 22-9, the LED should be operated only within the boundary established by the green line.

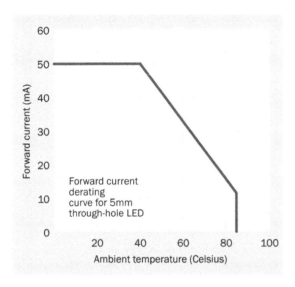

Figure 22-9. *Safe operation of an LED entails limiting the forward current if the temperature increases. The green line shows the boundary for operation of this particular component.*

View Angle

An LED formed from transparent epoxy or silicone (either water-clear or tinted) will create a well-defined beam with a view angle as narrow as 4 degrees or as wide as 160 degrees (in a few instances). The most common view angles for LED indicators are 30 degrees and 60 degrees.

Datasheets for LED indicators often include a *spatial distribution* graph showing the relative intensity of the light when viewed at various angles from the axis of the LED. The spatial distribution graph in Figure 22-10 is for an LED with a view angle defined as 40 degrees. This is the angle at

which the relative luminous intensity diminishes to 50%.

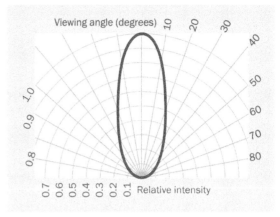

Figure 22-10. *A spatial distribution graph shows the relative intensity of light from an LED at various view angles.*

The view angle is of special concern in devices such as flashlights, where the spread of the beam affects the functionality.

How to Use It

Like all semiconductor devices, LEDs can be impaired by excess forward current and may break down irreversibly if subjected to excess reverse voltage. Their limits for reverse voltage are much lower than those of a rectifier diode. They are also vulnerable to heat, but are not particularly vulnerable to static electricity.

Polarity

A through-hole LED will have two leads of unequal length. The longer lead connects internally with the *anode* of the diode, and should be wired externally to the "more positive" side of a power source. The shorter lead connects internally with the *cathode* of the diode, and should be wired externally to the "more negative" side of a power source.

To remember the functions of the leads, consider that the plus sign would be twice as long as a minus sign if its horizontal and vertical stroke were disassembled and placed end to end.

If a round LED has a flange around its base, a flat spot in the flange will be closest to the cathode side of the component.

Series Resistor Value

Because the effective internal resistance of a diode is not a constant value at different voltages, a trial-and-error approach may be necessary to determine the ideal value for a series resistor with an LED indicator. For this purpose, a trimmer potentiometer can be used with a sample LED while measuring the current through it and the voltage drop across it. A fixed-value resistor can then be substituted. If the choice is between a resistor value that is a little too high and a value that is a little too low, the higher value resistor should be used.

An approximate value can be found using a very simple formula in which R is the resistor value, V_{CC} is the supply voltage, V_F is the forward voltage specified for the LED, and I is the desired current:

$$R = (V_{CC} - V_F) / I$$

Normally a series resistor rated at 1/4 watt will be acceptable, and 1/8 watt may be used in 5VDC circuits. However, care should be taken with a power supply of 9V or higher. Suppose an LED is rated for 1.8V forward voltage at 20mA. In a 5V circuit, the voltage drop across the series resistor will be:

$$V = 5 - 1.8 = 3.2$$

Therefore, the resistor must dissipate 3.2V * 20mA = 64mW. This is comfortably below the 125mW rating of a 1/8 watt resistor. However, with a 9V power supply, the voltage drop across the series resistor will be:

$$V = 9 - 1.8 = 7.2$$

Now the resistor must dissipate 7.2V * 20mA = 144mW. This exceeds the 125mW limit for a 1/8 watt resistor.

LEDs in Parallel

If multiple LEDs are to be driven in parallel, and none of them has to be switched individually, it is naturally tempting to save time by using a single series resistor for all of them. In these circumstances, assess the maximum current carefully and multiply by the voltage drop imposed by each of the LEDs, to determine the wattage of a series resistor.

Linking dissimilar LEDs in parallel is not recommended, because the threshold voltage decreases with increasing temperature. The hottest LED will therefore receive the largest current, and thus become even hotter. Thermal runaway can result.

LEDs containing their own series resistors can safely be wired in parallel.

Multiple Series LEDs

A series resistor wastes current by dissipating it as heat. In an application where two or more LED indicators will be illuminated simultaneously, the LEDs can be connected in series with a lower-value resistor, and three LEDs in series may eliminate the need for a resistor completely, depending on the voltage of the power supply. Here again a trimmer potentiometer should be used to determine an ideal value for any series resistor that may be necessary.

Comparisons with Other Light Emitters

Because LED indicators have largely replaced neon bulbs and miniature incandescent lamps, comparisons are of limited importance at this point. The situation regarding **LED area lighting** is different in that it is still competing actively with fluorescent lights and, in some instances, halogen. A list of advantages and disadvantages for high-intensity white LEDs is given in "Comparisons" on page 223. The advantages of incandescent lamps are listed in "Relative Advantages" on page 179.

Other Applications

LEDs are used in **optocouplers** and in **solid-state relays**. Usually an infrared LED is embedded inside a chip or a plastic module, and emits light through an interior channel to activate a phototransistor. This arrangement provides electrical isolation between the switching signal and the switched current.

Some sensors use an LED paired with a phototransistor at opposite sides of a U-shaped plastic mount. A sensor of this type can monitor industrial processes or may be found inside a photocopy machine, to detect the presence of a sheet of paper.

What Can Go Wrong

Excessive Forward Voltage

Like any diode, the LED has a *threshold voltage* in the forward direction. If this threshold is exceeded, the effective internal resistance of the LED falls very rapidly. Current rises equally rapidly, and quickly damages the component, unless it is protected by an appropriate series resistor.

Excessive Current and Heat

Exceeding the recommended value for forward current, or allowing an LED to overheat, will shorten its lifetime and cause a premature dimming of light output. LEDs generally require some current limiting or regulation (most commonly with a series resistor). They should not be connected directly to a voltage source such as a battery, even if the battery voltage matches the voltage of the diode. The exception to this rule is if the internal resistance of the battery is high enough to limit the current, as in the case of button-cell batteries.

Storage Issues

LEDs of different types are often indistinguishable from each other. They can also be indistinguishable from photodiodes and phototransistors. Careful storage is mandatory, and reusing LEDs that have been breadboarded may cause future problems if they are wrongly identified.

Polarity

If the leads on an LED indicator are trimmed, and if the indicator lacks a flange in which a flat spot will identify the cathode, the component is easily misused with reversed polarity. If it is connected with a component that has limited current sourcing capability (for instance, the output pin of a digital chip), the LED will probably survive this treatment. However, maximum reverse voltage is often as low as 5VDC. To minimize the risk of errors, the anode lead can be left slightly longer than the cathode lead when they are trimmed for insertion in a breadboard or perforated board.

Internal Resistors

As previously noted, it is difficult to distinguish an LED that contains its own series resistor from another LED that does not. The two types should be stored separately, and should be reused circumspectly.

LED area lighting

23

The term **LED area lighting** is used in this encyclopedia to describe a white LED source that is bright enough to illuminate rooms, offices, or outdoor areas. It may also be used in desk lamps or table lamps as *task lighting*. LEDs for these purposes may be categorized as *high-brightness*, *high-power*, *high-output*, or *high-intensity*. A complete fixture containing at least one light source is properly known as a *luminaire*, although the term is not uniformly applied and is sometimes written incorrectly as a *luminary*.

The full term *light-emitting diode* is not normally applied to an LED used for area lighting. For this purpose, the LED acronym has become universal. Periods are not normally placed between the letters.

While an LED area-lighting package may contain more than one diode, it is still categorized here as a single source. By contrast, any component consisting of multiple separately discernible light-emitting diodes, as in a seven-segment numeral, a 14- and 16-segment alphanumeric character, a dot-matrix character, or a display of multiple characters, is listed in a separate entry as an **LED display**.

The term *OLED* is an acronym for *Organic Light-Emitting Diode*, a thin panel in which an organic compound is contained between two flat electrodes. Despite its functionality as a form of LED, its design is similar to that of thin-film electroluminiscent light sources. Therefore it is discussed in the entry on **electroluminescence**.

OTHER RELATED COMPONENTS

- **LED indicator** (see Chapter 22)
- **incandescent lamp** (see Chapter 18)
- **fluorescent light** (see Chapter 20)
- **neon bulb** (see Chapter 19)
- **electroluminescence** (see Chapter 26)

What It Does

High-brightness white LEDs provide a plug-compatible alternative to **incandescent lamps**, *halogen lighting*, and **fluorescent lights** for work spaces and the home.

At the time of writing, products are still evolving rapidly in the field of LED area lighting. A shared goal of manufacturers is to increase efficiency while reducing retail price to the point where high-brightness LEDs will displace fluorescent tubes for most low-cost lighting applications.

A wall-mounted LED reflector-bulb that emulates a halogen fixture is shown in Figure 23-1. A small LED floodlight for exterior use is shown in Figure 23-2. An early attempt to package an LED area light in a traditional-style bulb is shown in Figure 23-3. Within a decade, as LED area lighting continues to evolve, some of these examples may look quaint. Configurations are evolving, with final results that remain to be seen.

Figure 23-1. *A small LED reflector-light emulating a halogen fixture. Note the square of yellow phosphors mounted on the diode.*

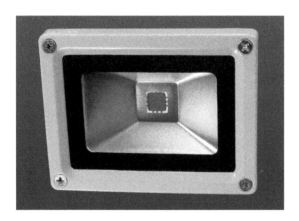

Figure 23-2. *A floodlight for exterior use. Nine LEDs are assembled behind the yellow phosphors. The steel frame measures about 4" by 3".*

Trends in Cost and Efficiency

The *luminous flux* of a source is the total power that it radiates in all directions, within the visible spectrum. The unit that measures luminous flux is the *lumen*. For a more detailed discussion of

this topic, see "Power" on page 177. Since 1965, the cost per lumen of light from a given color of LED has decreased by about a factor of 10, while the maximum number of lumens emitted by an LED package has increased by a factor of 20, during each decade. This is known as Haitz's Law, named after Dr. Roland Haitz of Agilent Technologies. Figure 23-4 illustrates it graphically.

Figure 23-3. *An LED light bulb. Unlike an incandescent bulb, it focuses the illumination in one direction, like a reflector-light. Consuming only 6W, it is claimed to be equivalent to a 40W incandescent bulb.*

Schematic Symbol

Schematic symbols that are commonly used to represent an LED are shown in Figure 23-5. The symbol remains the same regardless of the size or power of the component, but architectural plans may represent any type of light using the circle-and-X symbol at bottom right.

How It Works

A high-brightness LED functions on the same basis explained in the entry describing **LED indicators**. Photons are emitted when electrons are sufficiently energized to cross a PN junction and combine with electron-holes.

An LED that appears white, or off-white, actually emits blue light that is re-radiated over a wide range of wavelengths by adding a layer of yellow phosphors to the chip. A cutaway diagram of an

LED chip (properly known as a *die*), mounted under a silicone lens, is shown in Figure 23-6.

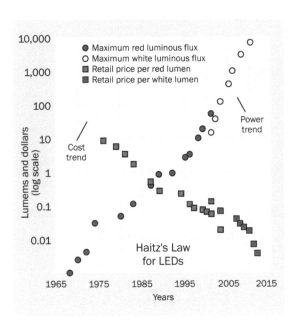

Figure 23-4. *The increase in light output (luminous flux, in lumens) of a single LED compared with the decrease in cost-per-lumen during the years since 1965. The vertical logarithmic scale measures both dollars and lumens. Source: Philips Gardco site-lighting fact sheet with additional data from a "Strategies in Light Report" published by Semiconductor Equipment and Materials International in 2013.*

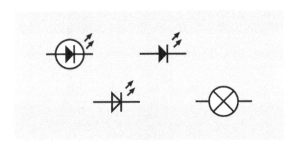

Figure 23-5. *The symbol for an LED remains the same regardless of its size and power, but architectural plans may use the circle-and-X symbol at bottom right for any type of light fixture.*

LEDs are mass-produced by etching them into crystals that are then cut into *wafers* before being subdivided into *dies*, like silicon chips. Most of the blue LEDs that form the basis of white lighting use sapphire crystals as their substrate. The crystal may range in diameter from two inches to six inches. Large sapphire wafers are also finding potential applications in camera lens covers and scratch-resistant cover plates for cellular phones.

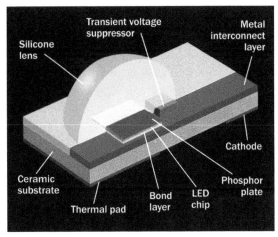

Figure 23-6. *Cutaway diagram of a high-brightness white LED. Adapted from Philips Lumileds Technical Reference document.*

While a die for an LED indicator may be 0.3mm x 0.3mm, a die in a high-brightness LED is often 1mm x 1mm. The size is limited by technical issues involving total internal reflection of the generated light.

The precise color of the light is adjusted by adding red phosphors to the yellow phosphors. This reduces the overall efficiency of the LED by around 10% but creates a "warmer" radiance. A graphical illustration of this principle is shown in Figure 22-3, in the entry on LED indicators.

The *color temperature* of white or offwhite light is measured in degrees Kelvin, typically ranging from 2,500K to 6,500K, where a lower number represents a light with more red in it and a higher number represents a light with more blue in it. This system of measurement was originally used with **incandescent bulbs** to define the temperature of the filament, which determined its color. See "Spectrum" on page 173 for a detailed explanation.

Visible Differences

The effects of different types of illumination are compared in Figure 23-7. To create this figure, first a color chart was prepared in Photoshop and printed on high-gloss photo-grade paper with a Canon Pro9000 Mark II inkjet printer, which has separate colors for red and green in addition to cyan, pale cyan, magenta, pale magenta, yellow, and black.

The color chart was then photographed twice with a Canon 5D Mark II, using a fixed white balance of 4000K. The first exposure was made with "daylight spectrum" LED lighting (claimed color temperature of 6500K) while the second was made with halogen lighting (claimed color temperature of 2900K). The photographs were adjusted in Photoshop for levels only, to fill the available range of 256 values. The two exposures show how the same chart would appear when viewed under the different lights, if the human eye did not adjust itself at all. Note the large area of the LED exposure which is rendered in shades of blue or purple. Also note the dullness of the reds. This confirms the everyday belief that "daylight spectrum" LEDs tend to have a cold, purplish cast while incandescents have a warmer, yellow look.

The same camera was then used to make two more exposures, this time with the white balance set to 6500 for LED lighting and 2900 for halogen lighting, which would be the recommended standard procedure, suggesting the kind of compensation that the human eye also tends to make for different ambient lighting. The result is shown at Figure 23-8. The LED version has improved, but the reds and yellows are still muted. The halogen version also looks better than before, but the magenta end of the spectrum has too much yellow in it. These images show the limits of white-balance correction for indoor photography.

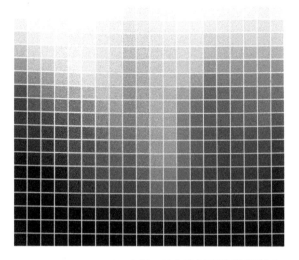

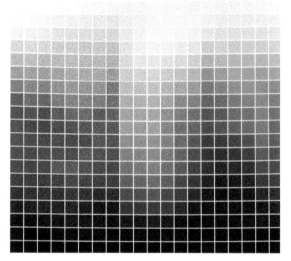

Figure 23-7. *The same printed color chart viewed with "daylight white" LEDs (top) and halogen lighting (bottom), without any compensation to allow for the different light spectra. A fixed white balance of 4000K was used for both pictures.*

Side-by-Side Comparison

Because the human eye is much better at comparing colors when they are adjacent to each other, another color chart was prepared using just six color bars of fully saturated red, yellow, green, cyan, blue, and magenta, with paler and darker versions above and below. The bars were separated with large white gaps. The chart was photographed first with the white balance set to

6500 under "daylight white" LED lighting and then again with the white balance at 2900 for halogen lighting. In Photoshop, the bars from the halogen version were copied and pasted beside the bars from the LED version to facilitate an A-B comparison. The result is shown in Figure 23-9.

figure shows the dramatic difference at the red end of the spectrum, and the poor reproduction of yellows by LED lighting. However, the LED rendering of green is better, and likewise the rendering of magenta, except where its darker version is concerned. Among the pale versions of the colors, the LED lights produce much less density (i.e., they have a brighter look) in the blues, greens, and cyans. The low densities will show up as pale highlights in a photograph of an object, and the picture will tend to have excessive contrast. This will also contribute to the "harsh" look of "daylight white" LED lighting which may be perceived by the eye.

Figure 23-9. *A range of six fully saturated colors, with lighter and darker shades added above and below, photographed first with "daylight spectrum" LEDs and then with halogen lights, after which the two sets of colors were paired for easy comparison. The LED version is on the left in each pair.*

Halogen is deficient at the blue-violet end of the spectrum, even when the camera has an appropriate white-balance setting. Photographers can correct this using image-editing software. LED "daylight spectrum" lights are more difficult to correct. LEDs classified as "warm" should reproduce reds better, but may not do so well with blues.

Diffuse light from a uniformly cloudy sky may be the most ideal form of lighting for photographing objects, but this is of little help for people who work (or take photographs) under artificial lights.

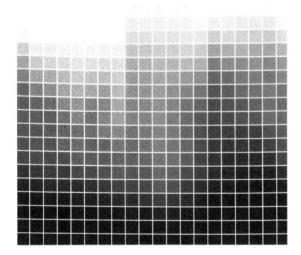

Figure 23-8. *The same color chart as before, photographed with appropriate color temperature settings of 6500K (top, using "daylight-spectrum" LEDs) and 2900K (bottom, using halogen).*

In each pair of colors, LEDs lit the one on the left, and halogen lights lit the one on the right. This

LED lights that contain separate red, green, and blue emitters may perform better, but create a different problem, in that shadows will tend to have color fringes caused by the small offsets between the colored emitters.

Heat Dissipation

An LED is less than 100% efficient because not all electrons mate with electron holes. Some manage to bypass the semiconductor junction; some recombine without generating light; and some transfer their energy to other atoms. In each instance, waste heat is created. While the heat in an incandescent bulb is mostly dissipated by radiation, an LED must get rid of the heat almost entirely by conduction, typically through a *heat sink*. This complicates the design of a fixture, because the integrity of the pathway to dispose of the heat must be retained when the LED bulb or tube is replaced.

Efficacy

The *radiant luminous efficacy* (LER) of an incandescent light source measures how effective it is at channeling its output within the visible spectrum instead of wasting it in infrared radiation. LER is expressed in lumens per watt, and in an incandescent bulb, it is calculated by dividing the power emitted in the visible spectrum (the *luminous flux*) by the power emitted over all wavelengths. This is described in detail in "Efficacy" on page 179 in the entry dealing with incandescent lamps.

In an LED indicator, almost all the radiation can be contained within the visible spectrum, which suggests that its efficacy should be 100%. However, because some waste heat is still created internally, the efficacy is calculated by dividing the light output, in lumens, by the power input, in watts, at the voltage required by the LED. (Lumens can be converted directly to watts, and therefore this division makes a comparison between similar units).

In an LED lighting fixture that contains its own electronics to convert higher voltage AC to lower voltage DC, the power consumption of the fixture is measured not at the diode, but at the input side of the electronics. Therefore, the inefficiency of the electronics reduces the efficacy value of the lighting unit.

Dimming

An incandescent bulb is very sensitive to reduction in power. It becomes radically inefficient, emitting perhaps 1% of its normal light output if the power is reduced to 40%.

LEDs have an almost linear response to the supplied power. Usually a triac-based dimmer will not work well with LED area lighting, and a dimmer designed for LEDs must be substituted, using pulse-width modulation.

Ultraviolet Output

The gas plasma in a fluorescent light generates ultraviolet wavelengths that are shifted to the visible spectrum by the phosphor coating inside the glass envelope. Imperfections in the phosphor coating can potentially allow leakage of ultraviolet light, causing some researchers to claim that the use of CFLs (compact fluorescent lighting) for close-up work with desk lamps can increase the risk of developing some forms of skin cancer. (This claim remains controversial.)

LED manufacturers are quick to point out that white LEDs do not emit any ultraviolet radiation. Figure 23-10 shows spectral power distribution curves derived from measurements of three high-brightness Color Kinetics LEDs manufactured by Philips. The manufacturer states categorically that "The LED-based color and white light products made by Color Kinetics do not emit outside the visible spectrum." Infrared radiation is also negligible.

Color Variation

The *correlated color temperature* (CCT) is determined by finding the conventional incandescent color temperature which looks most similar to the light from a white LED. Unfortunately, because the CCT standard is insufficiently precise, and because small manufacturing inaccuracies can occur, two LED sources with the same CCT number may still appear different when they are side by side. While the human eye adjusts itself to overall color temperature, it is sensitive to differences between adjacent sources. If two or more white LEDs in a lighting fixture do not have identical spectra, the difference will be noticeable.

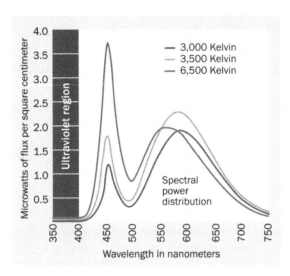

Figure 23-10. *Spectral power distribution curves for three high-brightness white LED lamps showing no ultraviolet emissions. (Adapted from a graph in a white paper published by Color Kinetics Incorporated.)*

To address the issue, manufacturers introduced the concept of "binning," in which lights are sub-classified to tighter specifications and are assigned bin numbers based on their measured characteristics. The Philips Optibin system, for instance, assesses the light from various angles, as well as perpendicularly to the source. This is especially important where a large area such as a building lobby is painted white and is lit by high-brightness LEDs that must appear uniform in color temperature.

Variants

LED area lighting products are often designed to emulate the form factors of incandescent bulbs,

halogen reflector bulbs, or fluorescent tubes. The standard screw-in base of an LED light bulb, the pin-base of a miniature 12V LED reflector bulb, and the pins on an LED tube enable easy migration to the newer technology.

Strip lights are unique to LED lighting systems. They are thick, flexible plastic ribbons in which are embedded a series of LEDs. For area lighting, the LEDs are white, and the strips can contain necessary control electronics for conversion of AC power. The strips can be placed behind ledges or moldings to provide soft, even illumination of the ceiling above.

Strip lights are also available for 12VDC power, to create lighting effects in customized automobiles and trucks. These strip lights are available in various colors in addition to white. Many have multicolor capability and can be controlled with a handheld remote.

Comparisons

The advantages of an **incandescent lamp** are listed in "Relative Advantages" on page 179, while advantages of **fluorescent lights** are listed in "Comparisons" on page 194. These lists can be compared with the following advantages for LED area lighting:

- While the life of an incandescent lamp for room lighting can be as little as 1,000 hours, LED area lighting typically claims up to 50,000 hours.

- The lifetime of an incandescent bulb is the average time it can emit light before catastrophic failure. The lifetime of an LED is the average time it can emit light before gradually dimming to 70% of its rated output. This is a much gentler, less inconvenient failure mode that does not require immediate replacement.

- Unlike a fluorescent light or incandescent bulb, the LED does not contain hot tungsten that fails as a result of erosion.

- Unlike a fluorescent light, an LED does not contain mercury, and therefore does not require special recycling arrangements that entail associated fees.

- While fluorescents can have difficulty starting in low temperatures, an LED is not sensitive to a cold environment.

- Bright LEDs are available in a wide range of colors that do not require filtering. Filters greatly reduce the efficiency of incandescent bulbs when they are used in applications such as traffic signals or rear lights on automobiles.

- High-brightness LEDs can be dimmable. Fluorescent lights are usually not dimmable, or perform poorly in this role.

- LEDs are inherently directional, because the die radiates light at an angle of 90 degrees to its plane. This makes it ideal for ceiling mounting, where as much light as possible should be directed downward. A fluorescent tube or incandescent bulb often requires a reflector which reduces the overall efficiency.

- LEDs are insensitive to cycling. The life expectancy of an incandescent bulb or (especially) a fluorescent tube is reduced by cycling it on and off.

- No flickering. Fluorescent tubes may start to flicker as they age.

- No electrical interference. Fluorescent tubes can interfere with AM radio reception and some audio devices.

- Safe from breakage. LED area lighting does not necessarily use any glass.

However, high-brightness LEDs still have some barriers to overcome:

- Cost. In the United States, before 60W incandescent bulbs were legislated out of existence, they could be sold profitably for less than $1 each. A T8 fluorescent tube, measuring 1″ diameter and 48″ long, currently costs between $5 and $6 (retail) but has a life expectancy in the region of 25,000 hours, and uses only 20% of the power of an incandescent bulb to generate two to three times as much light. Clearly the fluorescent tube is a more economical choice, despite the price of the electronics that must be included in the fixture to start the tube. By comparison, currently the purchase price of an LED tube is three times that of a fluorescent tube. It may last twice as long, but is not significantly more efficient, generating perhaps 100 lumens per watt while a fluorescent is typically capable of 90 lumens per watt. Prototype high-brightness LEDs have exceeded 200 lumens per watt, and should be competitive with fluorescents by 2020, but even then, migration will take time.

- Heat sensitivity. Heat reduces the light output and the lifespan of LED fixtures.

- Placement issues. Because LEDs are heat sensitive, they must be installed in locations that do not become excessively hot, their heat sinks must be correctly oriented, and they must have adequate ventilation.

- Color shift. Heat and age may cause the color temperature of an LED to shift slightly, as the color is usually derived from two types of phosphors.

- Nonuniformity. Manufacturing inconsistencies can cause LEDs of the same type to display slightly different color temperatures. Fluorescents and incandescents are more uniform.

- Lower heat output than incandescents. While this is an advantage from the point of view of efficiency, it can be a disadvantage in applications such as traffic signals or airport runway lighting where waste heat can help to keep the lights free from snow or ice.

Values

Although the output from an LED area light is directional, while the output from an incandescent bulb or a fluorescent light is omnidirectional, the intensity is measured the same way in each instance, using *lumens*. This unit expresses the total light emission, without taking directionality into account. (The intensity of **LED indicators** is calibrated in candelas, which measure the power within an angle of dispersion; but candelas are not used for area lighting.)

Typical values for incandescent bulbs are 450 lumens for a power consumption of 40 watts, 800 lumens for a consumption of 60 watts, 1,100 lumens for a consumption of 75 watts, and 1,600 lumens for a consumption of 100 watts. Because much of the output from an incandescent bulb may be wasted by using inefficient reflectors or allowing the light to shine in directions where it is not needed, a high-brightness LED rated at 1,000 lumens may actually appear brighter than a 75-watt incandescent bulb.

A T8 fluorescent tube measuring 48" long by 1" in diameter consumes only 32 watts but emits almost 3,000 lumens—when it is new. This output gradually diminishes by as much as 40% over the lifetime of the tube.

Incandescent bulbs deliver between 10 and 15 lumens per watt, approximately. A new fluorescent tube produces around 80 to 90 lumens per watt, and LED area lighting at the time of writing can provide 100 lumens per watt, under real-world conditions.

What Can Go Wrong

Wrong Voltage

Many high-brightness LED lighting units can be used with either 115VAC or 230VAC. There are exceptions, however. Check the specifications to make sure. Also, it is important to avoid applying domestic supply voltage to 12V LED miniature reflector-bulbs that are intended to replace 12VAC halogen bulbs of the same size.

Overheating

If a high-brightness LED fixture is equipped with a heat sink, this must be exposed to freely flowing air. Any vanes on the heat sink should be oriented vertically to encourage convection, and the fixture must not be placed in an enclosure. Overheating will radically shorten LED life.

Fluorescent Ballast Issues

A fluorescent fixture contains a *ballast* to limit the tendency of the tube to draw excessive current. The ballast is contained in a plastic box attached to the back of the frame in which the tube is mounted.

A *magnetic ballast* contains a coil, and is bypassed by an additional *starter* that applies unlimited current for one second when the power is switched on, preheating the tube to initiate plasma discharge.

An *electronic ballast* performs the same function without a separate starter.

Some LED tubes designed as substitutes for fluorescent tubes may allow a magnetic ballast to remain in the circuit, but may not tolerate an electronic ballast. Other LED tubes require any type of ballast to be unwired from the circuit. The unwiring operation will require disconnection of a couple of wires by removing wire nuts (assuming that the fixture has been designed to comply with U.S. building codes). The wires are then reconnected to apply power directly to the tube, and the wire nuts are reapplied to complete the new connection. The ballast can remain passively in the fixture.

Failing to remove the ballast and/or the starter from a fluorescent fixture before installing an LED tube that requires direct connection to the power supply can damage the tube. Connecting the power incorrectly to the LED tube may result in it failing to light up. Documentation supplied with the LED tube should provide guidance for disconnecting the ballast and connecting the tube. Note that the pin functions on LED tubes are not standardized at this time.

Misleading Color Representation

Because the spectrum of a white LED is not evenly weighted across all wavelengths, it will fail to represent some colors accurately, as shown previously. This can be important if LEDs are used to illuminate full-color printing or artwork, or if they are installed in stores selling merchandise such as clothes, furnishings, or food.

LED display

24.

In this encyclopedia, a component consisting of multiple separately discernible light-emitting diodes, such as a seven-segment numeral, 14- or 16-segment alphanumeric character, a dot-matrix character, or a display module containing multiple characters, is categorized as an **LED display**. The term *light-emitting diode* is hardly ever used to describe an LED display, as the LED acronym has become ubiquitous. The acronym does not usually include periods between the letters.

An **LED indicator** is defined here as a component usually 5mm or smaller in diameter, made of transparent or translucent epoxy or silicone, most often containing one *light-emitting diode*. It is purposed as a status indicator in a device, rather than as a source of illumination, and is sometimes referred to as a *standard LED*.

LEDs that are designed to illuminate large living or working areas are discussed in a separate entry as **LED area lighting**. They are sometimes referred to as *high-brightness LEDs* and almost always emit white light.

The term *OLED* is an acronym for *Organic Light-Emitting Diode*, a thin panel in which an organic compound is contained between two flat electrodes. Despite its functionality as a form of LED, its design is similar to that of thin-film electroluminiscent light sources. Therefore it is discussed in the entry on **electroluminescence**.

OTHER RELATED COMPONENTS

- **LED indicator** (see Chapter 22)
- **LED area lighting** (see Chapter 23)
- **vacuum-fluorescent** (see Chapter 25)
- **electroluminescence** (see Chapter 26)
- **LCD** (see Chapter 17)

What It Does

An LED display presents information on a panel or screen by using multiple segments that emit light in response to a DC current, almost always at a voltage ranging between 2VDC and 5VDC. The display may contain alphanumeric characters and/or symbols; simple geometrical shapes; dots; or pixels that constitute a *bitmap*.

A *liquid-crystal display*, or **LCD**, serves the same purpose as an LED display and may appear very similar, except that a liquid crystal reflects incident light while an LED emits light. The increasing use of backlighting with LCDs has made them appear more similar to LED displays.

There is no schematic symbol to represent an LED display. Where a segmented display is used, often the segments are represented with drawn outlines.

The simplest, most basic, and probably the best-known example of an LED display is the *seven-*

227

segment numeral, one of which is shown in Figure 24-1. This is a Kingbright HDSP-313E with a character height of 0.4″.

Figure 24-1. *The most basic LED display, able to create numerals from 0 through 9 using seven light-emitting segments that can be illuminated individually. An eighth segment forms the decimal point.*

How It Works

The process by which an LED generates light is explained in "How It Works" on page 207, in the entry dealing with **LED indicators**. Each light-emitting diode in an LED display is functionally the same as the diode in an LED indicator.

LEDs must be driven with DC. This is a primary distinction between an LED display and an LCD, which requires AC.

Variants

LCD comparisons

LCDs and LED displays can look very similar. This raises the obvious question: which is appropriate for a particular application?

LCDs (without backlighting) are more appropriate for applications such as digital watches and solar-powered calculators where power consumption must be minimized. They are capable of running for years from a single button cell.

LCDs are easily visible in bright ambient light, where LED displays are not. LCDs can also be designed to display complex pictographic shapes and symbols, while the segments of an LED display are more constrained to be simple in shape.

An LCD is more likely to be affected by temperature than an LED, and powering it entails some slight inconvenience, because it requires an AC source that is unlikely to be useful elsewhere in a circuit. If the LCD uses LED backlighting, it will also require a low-voltage DC power source for the backlight. An LED display is easier to use in that it can be driven directly from a microcontroller or logic chip, with only some series resistors to limit the current, and the addition of transistors to provide additional power where necessary.

Seven-Segment Displays

Early seven-segment LED displays were used in digital calculators, before LCDs became an affordable, practical alternative that greatly extended battery life. Initially, the size of the diodes was limited, sometimes requiring magnifying lenses to make them legible.

Seven-segment displays are still used in some low-cost applications, although LCDs have become more common.

Figure 24-2 shows how the segments are identified with letters *a* through *g*. This scheme is used universally in datasheets, and is also used for LCDs. The decimal point, customarily referred to as "dp," is omitted from some displays. The segments are slanted forward to enable more acceptable reproduction of the diagonal stroke in numeral 7.

Although seven-segment displays are not elegant in appearance, they are functional and are reasonably easy to read. They also enable the representation of hexadecimal numbers using letters A, B, C, D, E, and F (displayed as A, b, c, d, E, F because of the restrictions imposed by the small number of segments), as shown in Figure 24-3.

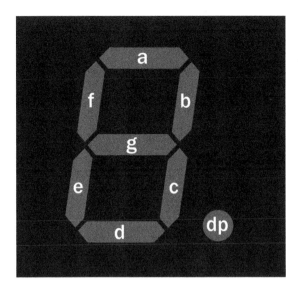

Figure 24-2. *A seven-segment LED display. The lower-case identifying letters are universally used in datasheets.*

In appliances such as microwave ovens, very basic text messages can be displayed to the user within the limitations of seven-segment displays, as suggested in Figure 24-4.

Numbers 0, 1, and 5 cannot be distinguished from letters O, I, and S, while letters containing diagonal strokes, such as K, M, N, V, W, X, and Z, cannot be displayed at all.

Multiple Numerals
Displays consisting of a single numeral are now rare, as few applications require only one digit. Displays of two, three, and four digits are more common, as shown in Figure 24-5.

Additional Segments
Displays with 14 or 16 segments were introduced in an effort to enable the representation of all the letters of the alphabet. The segment layout of these LED displays is identical to that of comparable LCDs. The differences between 14-segment and 16-segment displays are shown in Figure 24-6. Some are angled forward like seven-segment displays, even though the addition of diagonal segments makes this unnecessary for display of characters such as numeral 7.

Figure 24-3. *Numerals and the first six letters of the alphabet created with seven-segment displays.*

Figure 24-7 shows the scheme for identifying the segments of a 16-segment display. This naming convention is used in all datasheets. The lowercase letters that were customary with seven-segment displays are usually abandoned in favor of uppercase, perhaps to avoid confusion with the letter L. Note that letter I is omitted from the sequence.

For a complete alphanumeric character set enabled by a 16-segment display, see Figure 17-9 in the entry discussing LCDs.

An example of a 16-segment alphanumeric LED display is shown in Figure 24-8, mounted on a breadboard and wired to show the letter N. This is a Lumex LDS-F8002RI with a character height of 0.8". The component is still available at the time of writing, but in limited quantities.

Generally speaking, 16-segment displays were never very popular, because the gaps between adjacent segments impaired legibility. LED versions remain more readily available than LCD versions, but dot-matrix displays allow a better-looking, more easily legible alphabet, with the added possibility of simple graphics.

Dot-Matrix Displays

In the 1980s, some personal computers used a video character set in which each letter, numeral, punctuation mark, and special character was formed on a video screen from a fixed-size matrix of dots. A similar alphabet is now used in LED dot-matrix displays (and LCDs, as shown in Figure 17-10).

Figure 24-5. *Multiple seven-segment LED displays are often combined in a single component. Top: An Avago 2.05VDC 20mA display designed for a clock. Bottom: A Kingbright two-digit display which draws 20mA at 2.1VDC. The unlit outlines of the numerals would normally be hidden behind panels that are tinted to the same colors emitted by the LED segments when lit.*

Figure 24-4. *Basic text messages can be generated with seven-segment displays, although they cannot represent alphabetical letters containing diagonal strokes.*

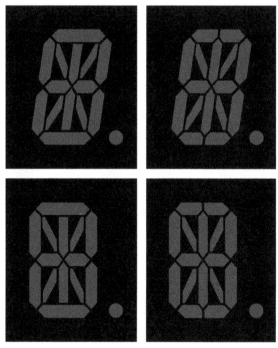

Figure 24-6. *Layouts for 14-segment and 16-segment alphanumeric LEDs are identical to those of LCDs.*

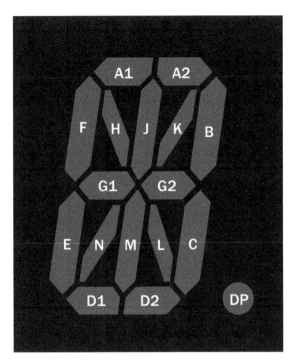

Figure 24-7. *The scheme for identification of segments in a 16-segment alphanumeric LED display.*

Figure 24-8. *A 16-segment alphanumeric LED display showing the letter N.*

Alphanumeric dot-matrix characters are often grouped in two or more rows with eight or more characters per row. The number of characters is always listed before the number of rows, so that an 8x2 display would contain eight alphanumeric characters in two horizontal rows. This type of component is properly described as a *display module*.

Display modules are used in consumer electronics products such as a stereo receiver where simple status messages and prompts are necessary —for example, to show the tone control settings or the frequency of a radio station. Because the cost of small, full-color, high-resolution LCD screens has been driven down rapidly by the mass production of cellular phones, and because these high-resolution screens are much more versatile, they have already displaced dot-matrix display modules in many automobiles and are likely to follow a similar path in other devices.

Pixel Arrays

The 8x8 pixel array of LED dots shown in Figure 24-9 measures 60mm square (slightly more than 2") and contains 64 LEDs, each approximately 5mm in diameter. Similar arrays are available in other sizes and with different numbers of dots. Displays of the same type may be assembled edge-to-edge to enable scrolling text or simple graphics.

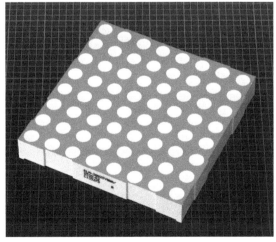

Figure 24-9. *An 8x8 matrix of LED dots measuring more than two inches square.*

Multiple Bar Display

A bar display is a row of small rectangular LEDs in a single component. It may be used for digital representation of an analog signal. The higher the voltage of the signal, the more bars will be illuminated. A typical application would be to show the signal strength of an input to an audio recorder. Ten bars are often used, as in the display shown in Figure 24-10, but multiple components can be combined end to end.

Figure 24-10. *Two LED bar displays in which segments can be lit individually.*

Single Light Bar

A light bar can be thought of as a single-source LED, as it is configured as a single square or rectangle. It is mentioned here, rather than in the entry for single-source LED indicators, because variants may be subdivided into two, three, four, or (sometimes) more discrete sections. These variants are often included in the same datasheet as the monolithic version.

A light bar contains multiple LEDs (often, four in number) behind a translucent panel that provides evenly diffused radiance.

Values

The values for most LED displays are basically the same as for LED indicators, in terms of color,

brightness, current consumption, and voltage. See "Values" on page 211 for information.

Multiple-character dot-matrix LED display modules may have different requirements for forward voltage and forward current, depending on drivers that are incorporated in the module. Because there is no standardization for these modules, it will be necessary to consult the manufacturer's datasheet.

How to Use It

Seven-Segment Basics

The diodes in a seven-segment LED display share either a *common anode* or a *common cathode*, the latter being more frequently used. The two types of internal wiring are provided for convenience only. Externally, the displays function identically.

A schematic suggesting the internal wiring and pinouts of a typical ten-pin common-cathode display is shown in Figure 24-11. The pins are numbered as seen from above. Appended to each number is the identity of the segment to which it is connected. Pins 3 and 8 are connected with the cathodes of all the internal LEDs. Both of these pins should be used, to serve as heat sinks for the display.

Note that series resistors are not included inside the display and must be added externally. Their value will be determined by the power supply, to limit the forward current and forward voltage through the LEDs to the extent specified by the manufacturer.

An encapsulated *resistor array* containing either seven or eight resistors in an SIP or DIP chip can be used instead of individual resistors. A seven-segment LED display would require the type of resistor array in which both ends of each resistor are accessible.

Where two or more numerals are combined in a single component, this type of display is likely to have two horizontal lines of pins. In this case, pin 1 will be at the bottom-left corner, seen from above. As always, the pins are numbered counterclockwise, seen from above.

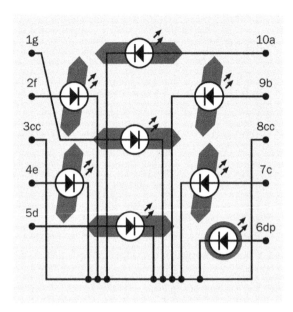

Figure 24-11. *A schematic view of internal connections and pinouts of a seven-segment common-cathode LED display. The numbers refer to the pins of the component, seen from above. The 1 pin may also have a mark beside it on the component, for identification. The orientation of the display can be deduced if there is a decimal point, as this should be at the bottom-right corner.*

Where three or more numerals are combined in a single component, the pinouts may be designed for multiplexing rather than individual access to every segment of each numeral. A four-digit clock display, for instance, may have seven pins that connect in parallel to respective segments in all of the numerals, and four additional pins that can ground each numeral in turn, so that they can be selected sequentially.

Driver Chips and Multiplexing

Illuminating the appropriate segments in a single numeral can be done directly from a microcontroller, or through a driver chip such as the well-known and widely used 4543B that converts a binary-coded decimal input into appropriate segment output patterns. The chip can source sufficient current to drive each segment through a series resistor. Its pinouts are shown at Figure 24-12.

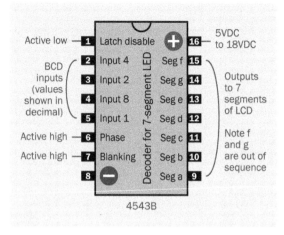

Figure 24-12. *Pinouts of the 4543B seven-segment LED driver chip.*

When used in conjunction with a microcontroller, the 4543B can drive several seven-segment displays by multiplexing them. The basic schematic to achieve this is shown in Figure 24-13, omitting optional features such as leading-zero blanking or connections for a decimal point. The microcontroller sends the binary code for the first numeral and simultaneously grounds the common cathode of that numeral through a transistor, which is needed because as many as seven segments of the numeral may be passing current in parallel. The microcontroller then sends the binary code for the second numeral, and grounds it; then sends the binary code for the third numeral, and grounds it; and the cycle repeats. So long as this process is performed at sufficient speed (at least 50Hz), persistence of vision will create the illusion that all the numerals are active simultaneously. The circuit can be compared with a similar circuit to drive LCDs, shown in Figure 17-17.

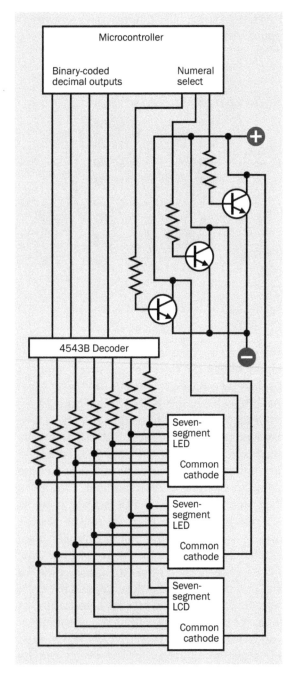

Figure 24-13. *A basic, simplified circuit for driving multiple seven-segment LED displays by multiplexing them.*

The disadvantage of this system is that the microcontroller must update the numerals constantly while performing other duties. To reduce this burden, a "smarter" driver such as the

MC14489 can be used, controlling up to five 7-segment digits, or the Intersil ICM7218, which can control up to eight 7-segment digits.

The MC14489 controller receives data serially, using SPI protocol, and handles the details of addressing the LEDs. Because it contains latches to sustain the displayed data, a microcontroller only needs to communicate with the driver when the displayed information needs to be updated.

The ICM7218 is a more sophisticated chip, available in several variants, one of which can receive data on an 8-bit bus and run the seven-segment displays in hexadecimal mode.

Sixteen-Segment Driver Chip

The MAX6954 by Maxim can drive up to eight 16-segment alphanumeric LED displays using a scheme known as Charlieplexing, named after a Maxim employee named Charlie Allen who came up with the concept as a way of reducing the pin count required for multiplexing. Other Maxim controllers use this same protocol, which is transparent to the user.

A microcontroller sends data serially via I2C protocol to the MAX6954, which contains a variety of features. It can drive 14-segment and 7-segment displays as well as 16-segment displays, and contains a 104-character alphabet for each of them. Setting up a microcontroller to send the various necessary command codes to the MAX6954 is not a trivial matter, and bearing in mind the probably impending end-of-life of 16-segment displays, a better option may be to use dot-matrix LED display modules that have controller logic built in.

Dot-Matrix LED Display Modules

A dot-matrix LED display module requires data to define a character set, and a command interpreter to process instructions that will be embedded in a serial data stream. These capabilities are provided either by separate chips or (more often) are incorporated into the LED display module itself.

The SSD1306 is a monochrome graphical controller capable of I2C or SPI serial communication, or parallel communication. When this capability is built into a display module, only one of these types of communication may be activated.

The SSD1331 is a color graphical controller with similar communication capabilities.

The WS0010 is a monochrome controller, compatible with HD44780, which is designed to control LCDs.

Typical controller functions are summarized in "Alphanumeric Display Module" on page 168. Because there is no standardization in this field, precise details must be found in manufacturers' datasheets.

Pixel Arrays

The connections inside an 8x8 pixel array are shown in Figure 24-14, where the schematic symbols for LEDs have been replaced by gray circles for space reasons. To illuminate one LED, power is supplied to the intersection where it resides. In the figure, each vertical conductor (identified as A1, A2 ... A8) can power the anodes of a column of eight LEDs, while each horizontal conductor (identified as C1, C2 ... C8) can ground the cathodes of a row of eight LEDs. If only one vertical conductor is connected with positive power while one horizontal conductor is grounded, only one LED will light up, at the intersection of the active conductors.

A problem occurs if we wish to illuminate two LEDs. Suppose they are located at (A3,C2) and (A6,C5). Unfortunately, providing power to them will also result in activating LEDs at (A3,C5) and (A6,C2), as shown in Figure 24-15, where the yellow circles represent LEDs that have been switched on.

The answer to this problem is to rasterize the process. In other words, data is supplied on the array one line at a time, as in the process by which a TV picture is generated. If this is done quickly enough, persistence of vision will create the illusion that the LEDs are illuminated simultaneously.

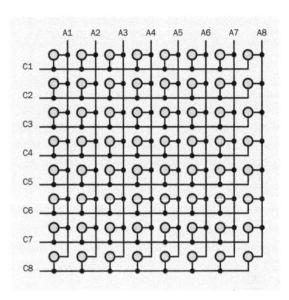

Figure 24-14. *Internal connections in the 8x8 matrix. Each gray circle represents an LED.*

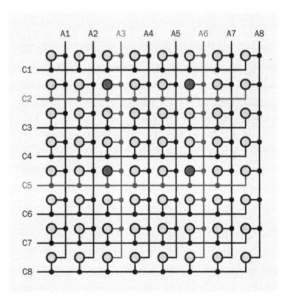

Figure 24-15. *An attempt to illuminate LEDs at (A3,C2) and (A6,C5) will also activate the LEDs at (A3,C5) and (A6,C2).*

A form of multiplexing is used to achieve this. One row of LEDs is connected to negative ground for a brief interval. During this interval, the anodes of selected LEDs are powered momentarily. Then the next row is grounded, and selected LEDs along that row are powered momentarily. This process is repeated for all eight rows before being repeated.

If several 8x8 matrices are assembled edge to edge, their horizontal conductors can be common to all of them. A horizontally scrollable display (sometimes referred to by the archaic term, *electric newspaper*) would then be possible, although the circuit design would be nontrivial.

Multiple Bar Display Driver

The LM3914 is a driver for a bar display that compares an analog input with a reference voltage and provides power to the segments of a multiple bar display, ranging from 2mA to 30mA, adjustable to match the specification of the display that is being used. The chip can generate either a "thermometer" effect, as more outputs are activated when the analog input increases, or a "moving dot" effect, in which only one output is on at a time.

One-Digit Hexadecimal Dot Matrix

While multi-character dot-matrix LED display modules are a versatile way to display prompts and numbers, a simpler component is sometimes sufficient. The Texas Instruments TIL311 is a minimal dot-matrix LED display that receives a binary value from 0000 through 1111 on its four input pins and generates the output in hexadecimal form, using numerals 0 through 9 and letters A through F. The sixteen possible outputs in the dot-matrix display are shown in Figure 24-16. Although this component is no longer being manufactured, it is widely available from many sources, especially in Asia. It eliminates the series resistors and controller chip that are customary for a seven-segment display, and has a better-looking output.

Figure 24-16. *Sixteen possible outputs that can be displayed by the Texas Instruments TIL311 to show a hexadecimal value in response to a four-digit binary input.*

A sample of the TIL311 is shown in Figure 24-17 displaying the number 2.

If two or more of these chips are put together, they can be multiplexed to display multi-digit decimal or hexadecimal integers.

The chip features two decimal points, one to the left of the displayed numeral, and one to the right. If they are activated, they require their own series resistors to limit the current.

Figure 24-17. The Texas Instruments TIL311 can be driven directly by a microcontroller or counter chip, with no series resistors necessary. It can generate a hexadecimal output.

What Can Go Wrong

Common Anode versus Common Cathode

An LED display containing a common cathode is usually identical in appearance to a display containing a common anode, and the two versions will be distinguished by only one digit or letter in their part numbers. Because LED displays have a limited tolerance for reverse voltage, part numbers should be double-checked before applying power.

Incorrect Series Resistance

A common error is to assume that only one series resistor is necessary for a seven-segment LED display, either between the common cathode pin and ground, or, if there is a common anode, between that and the positive power supply. The problem is that if the resistor is suitable for a single LED, its value will be too high when several segments of the display are sinking current or drawing current through it. If its value is reduced, it will be too low when only two segments are using it (as when generating the number 1).

To provide equal illumination of all the segments, each must have its own series resistor.

Multiplexing Issues

When several displays are multiplexed, they naturally appear dimmer, creating a temptation to compensate by upping the current. Because current is only being applied to each display intermittently, a natural assumption is that a higher current can be safely used.

This may or may not be true. When running an LED device with pulsed current, the peak junction temperature, not the average junction temperature, determines the performance. At refresh rates below 1kHz, the peak junction temperature is higher than the average junction temperature, and the average current must therefore be reduced.

Datasheets must be checked to determine whether a device is designed with multiplexing in mind and, if so, what the recommended peak current is. Very often this value will be accompanied by a maximum duration in milliseconds, and a calculation may be necessary to determine the refresh rate, bearing in mind how many other LED displays are being multiplexed in the same circuit at the same time.

Irresponsible multiplexing will shorten the life of an LED display or burn it out.

vacuum-fluorescent display

25

The term **vacuum-fluorescent display** is seldom hyphenated, but the first two words are hyphenated here as they constitute an adjectival phrase. The acronym *VFD* is becoming increasingly popular, although it is ambiguous, being also used to identify a *variable frequency drive*. In both instances, the acronym is printed without periods between the letters.

The entry in this encyclopedia dealing with **fluorescent lights** does not include VFDs, because their purpose and design are very different. A VFD is an informational display, often showing numerals and letters, while a fluorescent light merely illuminates a room or work area. Although a VFD does use fluorescent phosphors, they are printed onto light-emitting segments of the display instead of being applied to the inside surfaces of a glass envelope.

OTHER RELATED COMPONENTS

- **LED indicator** (see Chapter 22)
- **LCD display** (see Chapter 17)
- **electroluminescence** (see Chapter 26)

What It Does

A **vacuum-fluorescent display** or *VFD* superficially resembles a backlit monochrome **LCD** or an **LED display**, as it can represent alphanumeric characters by using *segments* or a *dot matrix*, and can also display simple shapes. It is often brighter than the other information display systems, and can emit an intense green phosphorescent glow that some people find aesthetically pleasing, even though a grid of very fine wires is superimposed internally over the displayed image.

There is no specific schematic symbol to represent a vacuum-fluorescent display.

How It Works

The display is mounted inside a sealed capsule containing a high vacuum. A widely spaced series of very fine wires, primarily made of tungsten, functions as a *cathode*, moderately heated to encourage electron emission. The wires are often referred to as *filaments*.

A **fluorescent light** uses AC, and both of its electrodes are often confusingly referred to as cathodes. A VFD uses DC, and its cathode array has the function that one would expect, being connected with the negative side of the DC power supply.

Opposite the cathode, just a few millimeters away, is an anode that is subdivided into visible alphanumeric segments, symbols, or dots in a matrix. Each segment of the anode is coated with *phosphors*, and individual segments can be separately energized via a substrate. When electrons strike a positively charged anode segment, it emits visible light in a process of *fluorescence*. This behavior can be compared with that of a *cathode-ray tube*. However, the cathodes in a VFD are efficient electron emitters at a relatively low temperature, while the cathodes in a cathode-ray tube require substantial heaters.

Anode, Cathode, and Grid

A *grid* consisting of a mesh of very fine wires is mounted in the thin gap between the filaments of the cathode and the segments of the anode. A simplified view of this arrangement is shown in Figure 25-1.

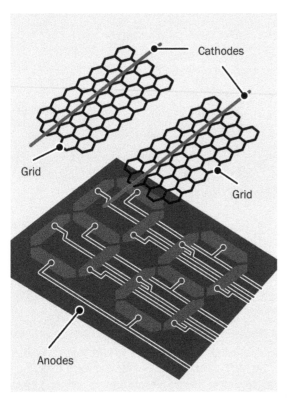

Figure 25-1. *The basic elements of a vacuum-fluorescent display.*

The polarity of the charge on the grid controls and diffuses electrons emitted by the cathode. If a grid section is negatively charged, it repels electrons and prevents them from reaching the

sections of the anode beneath it. If the grid section is positively charged, it encourages electrons to reach the anode. Thus, the grid functions in the same way as the grid in a triode vacuum tube, but its conductors are so thin, they are barely visible.

How to Use It

Electronic calculators used vacuum-fluorescent displays during the 1970s, before LED displays and LCDs became more competitive. Purely numeric VFD modules are still available as strings of digits, although they are becoming uncommon and have been replaced by alphanumeric dot-matrix modules where each VFD character is mounted in its own glass module on a separate substrate.

Figure 25-2 shows the interior of a Commodore calculator from the 1970s, with its nine-digit vacuum-fluorescent display enclosed in one glass capsule.

Figure 25-2. *The vacuum-fluorescent display from a 1970s Commodore calculator.*

A closeup of three digits from the previous figure appears in Figure 25-3, showing the grid superimposed above each numeral.

Figure 25-3. *Three digits from the previous figure, showing the grid that controls their illumination.*

A tinted filter of the same color as the display is usually placed in front of it, to conceal some of its workings. Thus, the Commodore calculator used a green filter in front of its green numerals. Figure 25-4 shows two seven-segment numerals from another device, with the filter removed. This reveals not only the grid but also the horizontal wires that function as the cathode. Connections between the segments of the numerals and a backplane are also visible.

Figure 25-4. *Seven-segment numerals viewed without a colored filter, revealing the cathode (horizontal wires) and the grid (wire mesh).*

Modern Application

A modern VFD module is likely to be mated with a driver that converts 5VDC to the higher voltage (typically 50VDC to 60VDC) required for the display. Built-in logic may offer the option to receive data via an 8-bit parallel bus or with SPI serial protocol, and will contain a character set. A typical display resolution is 128 x 64 pixels.

The combination of a grid and a segmented anode enables a VFD to be controlled by *multiplexing*. For instance, in a display of four seven-segment numerals, the same equivalent segments in all four numerals can be connected in parallel while a separate grid covers each numeral. When each grid is positively energized, it selects the corresponding numeral, and the on-off segment patterns appropriate to that numeral are supplied. This procedure is repeated for each numeral in turn. Persistence of vision makes it appear that they are all active simultaneously.

Variants

Color

Although a VFD cannot provide a full-color display, selected anode segments can be coated with different phosphor colors, which can fluoresce simultaneously. Two or three individual colors are typically used, as in the display for a CD player where color helps to distinguish a variety of different functions. A closeup of a portion of the display from a CD player (with color filter removed) appears in Figure 25-5.

Figure 25-5. *The lefthand section of a vacuum-fluorescent display from a CD player.*

Character Sets and Pictorial Design

In the past, VFDs have combined seven-segment numerals in the same display as custom-shaped anodes. Solid-state gain meters in an audio amplifier, for instance, have used numerals beside a pictorial representation of gain levels resembling analog meters. The look and layout of a display of this type has been unique to a particular product.

Modern VFDs tend to use a generic dot-matrix display in which a character set in firmware dictates how patterns of dots are grouped to form numbers, letters, symbols, or icons.

The appearance of character sets generated with generic segments and dot-matrix arrays is thoroughly discussed and illustrated in the entry describing liquid-crystal displays in Chapter 17. VFD alphanumeric modules are identical in visual design to LCD modules, even though the internal electronics are different.

Comparisons

Two advantages of a VFD are that it functions well at low temperatures (unlike an LCD) and has sufficient brightness and contrast to be usable in sunlight (unlike many LED displays). It can be viewed from almost any angle.

Typical applications have included digital instrumentation in automobiles, information displays in audio and video consumer-electronics equipment, and numerical readouts in vending machines, medical devices, and some digital clocks.

Because a VFD requires a relatively high voltage, has significant power consumption, can show only a limited range of fixed colors, and is more expensive than LED displays or LCDs, its popularity has declined since the end of the 1990s.

What Can Go Wrong

Fading

VFDs gradually fade with age, as a result of reduced electron emission from the electrodes or diminishing performance of the phosphor coatings. Increasing the working voltage can prolong the life of a display.

electroluminescence

The field of **electroluminescent** devices is sometimes referred to as *EL*. The same acronym can also be applied as an adjective to an individual electroluminescent device, as in, for example, "an EL panel."

An *organic light-emitting diode*, more commonly known by its acronym *OLED*, is included in this entry because it is technically an electroluminescent device and its design concept is similar to that of an electroluminescent panel. Generic LEDs are also technically electroluminescent, but are not commonly described as such, and have their own entries in this encyclopedia under the subject categories **LED indicator**, **LED area lighting**, and **LED display**.

OTHER RELATED COMPONENTS

- **LED indicator** (see Chapter 22)
- **LCD display** (see Chapter 17)
- **fluorescent light** (see Chapter 20)
- **vacuum-fluorescent** display (see Chapter 25)

What It Does

An electroluminescent device configured as a *panel*, *ribbon*, or *rope-light* contains phosphors that emit light in response to a flow of electricity.

Panels can be used as backlights for LCD displays or, more often, as always-on low-power devices such as exit signs and night lights. Ribbons and rope lights (the latter being also known, more accurately, as *light wires*) are used mainly as recreational novelties. They can be battery powered through a suitable voltage converter. A battery-powered rope light can be wearable.

Thin-film OLED electroluminescent panels are used in small video screens in handheld devices. At the time of writing, OLED TV screens measuring 50" or more have been demonstrated, but are not yet economic for mass production.

No specific schematic symbol exists to represent any electroluminescent device or component.

How It Works

Luminescence is the emission of light as a result of a process that does not require heat. (The opposite phenomenon is *incandescence*, in which heating causes an object to emit light; see Chapter 18 for a description of **incandescent lamps**.)

Electroluminescence is luminescence resulting from stimulation by electricity. This very broad definition really includes devices such as **LED**s, although they are hardly ever described in those terms. Electroluminescence generally refers to panels, films, or wires where electrodes are in direct contact with light emitters such as phosphors.

The exception is an *organic LED*, usually known by its acronym *OLED*, which is frequently described as an electroluminescent device, perhaps because its configuration as a sandwich of thin, flat layers resembles an electroluminescent panel. Two of the layers are semiconductors, and they interact as light-emitting diodes.

Phosphors

A *phosphor* is a compound such as zinc sulfide that will emit light when it receives an energy input from another light source or from electricity. Typically the compound must be mixed with an *activator* such as copper or silver.

For many decades, TV sets and video monitors were built around *cathode-ray tubes* in which the interior of the screen, at the front of the tube, was coated with phosphors. A beam of electrons that fluctuated in intensity generated a picture on the screen by drawing it as a series of lines.

Derivation

The term *phosphor* is derived from *phosphorescence*, which in turn comes from the name of the element *phosphorous*, which will glow when it oxidizes in moist air. (These terms were established before other forms of luminescence were discovered and understood. The behavior of phosphorous is really an example of *chemiluminscence*.)

For our purposes, a phosphor is a compound that is capable of fluorescence or electroluminescence.

Variants

Panels

Electroluminescent panels using phosphor powder, sometimes referred to as *thick phosphor*, are a popular choice where a constant, uniform, low light output is acceptable.

An electric potential is established between two films that act as electrodes, separated by a layer of phosphor crystals. Some manufacturers refer to this configuration as a *light-emitting capacitor* because the structure resembles a capacitor,

even though that is not its purpose. The front film is transparent, allowing light to escape.

An electroluminescent panel can be powered by AC or DC but requires at least 75V. Its power consumption is self-limiting, so that no control electronics are required other than a voltage converter if battery power is used.

The phosphors generate a constant, evenly distributed luminescence over the entire area, although the output is not very intense. Applications include night-lights, exit signs, and backlighting for wristwatches.

Panelescent electroluminescent lighting by Sylvania was used for instrument panel displays in some car models such as the Chrysler Saratoga (1960 through 1963) and Dodge Charger (1966 through 1967). It is still used for night-lights. *Indiglo* electroluminiscent displays are still widely used in wristwatches.

The interior components of a disassembled electroluminescent night-light are shown in Figure 26-1. The panel emits a natural pale green glow. A separate blue or green filter passes the glow while blocking other colors of incident light that would otherwise reflect off the panel.

Figure 26-1. *The two interior components of an electroluminscent night-light: the luminescent panel, and a separate translucent filter.*

Electroluminescent night-lights were popular in the 1970s and 1980s, often featuring cartoon characters to appeal to children. Figures 26-2 and 26-3 show the same night-light in its daytime off-state and its night-time on-state, respectively.

Figure 26-2. *A vintage Panelescent brand night-light, several decades old, in its off-state.*

Figure 26-3. *The same night-light, with its green radiance visible under conditions of low ambient light.*

Advantages of electroluminescent panels include the following:

- Low current consumption. One US manufacturer claims that a single exit sign will use electricity costing less than 20 cents per year, while the annual cost of a night-light will be less than 3 cents per year.

- Long life, up to 50,000 hours.

- Self-regulating; no control circuitry required.

- Omnidirectional light output.

- Very wide operational temperature range, between approximately -60 and +90 degrees Celsius.

- Can be plugged directly into a wall outlet.

Disadvantages include:

- Limited light output.

- Very limited choice of colors.

- Not very efficient, 2 to 6 lumens per watt (although the low light output naturally entails low power consumption).

- Gradual reduction in phosphor performance over time.

- High voltage required: 60V to 600V. Ideal for plugging into a wall outlet, but requires a converter when used with battery-powered devices.

Flexible Ribbons

The light-emitting layers inside a night-light are somewhat flexible, and can be made more flexible by reducing their thickness. The result is an electroluminescent ribbon that has some novelty value, and may be used for customizing automobiles. Figure 26-4 shows a ribbon about 1.5" wide and 12" long, designed for 12VDC power applied through an inverter.

Rope Light

A rope light or wire light may resemble a *glow-stick*. However, a glowstick generates light from chemiluminescence (chemical reactions that release photons), while a rope light uses electricity.

Figure 26-5 shows a rope light powered by two AA batteries connected through an inverter.

At the center of the rope light is a conductor that serves as one electrode. It is coated in phosphors, and the layer of phosphors is protected by a transparent sheath. One or more thin wires is wrapped around the sheath in a spiral, with large gaps between one turn and the next. These wires serve as the second electrode. The wires are enclosed in transparent insulation that forms an outer sheath.

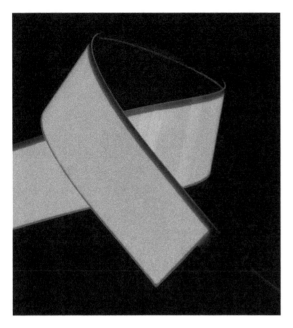

Figure 26-4. *A 12" length of electroluminescent ribbon.*

When AC is applied between the electrodes, the layer of phosphors emits light that radiates out in the gaps between the thin wires. The color of the light can be modified by using tinted outer insulation.

OLED

An OLED uses two thin, flat electrodes, somewhat like a thick-phosphor electroluminescent panel, except that it contains more layers and is capable of generating more light. The layers in an OLED are "organic" in that they consist of chemically organic molecules containing carbon and hydrogen atoms and generally do not contain heavy metals.

While an LCD video monitor or TV screen must have a separate backlight, an OLED generates its own light. This reduces the thickness of the display to a few millimeters and makes it potentially more efficient.

The semiconductor layers are subdivided into pixels, each functioning as a light-emitting diode, while additional layers carry a matrix of conductors for pixel addressing. In an AMOLED, the conductors form an active matrix, while in a PMOLED, they form a passive matrix.

Figure 26-5. *A length of glowing rope light, also known as a light wire.*

In an active matrix, each pixel is backed with a *thin-film transistor* to store its state while the energizing voltage transitions. This is often described as a *TFT* display; but the term is interchangeable with "active matrix."

In a passive matrix, each pair of conductors simply supplies current to a pixel. This is cheaper and easier to fabricate but is less responsive.

The terms "active matrix" and "passive matrix" have the same meaning as when used to describe a **liquid-crystal display**.

Monochrome OLED display modules with dot-matrix characters are currently available from China for just a few dollars. Although they appear superficially similar to LCD modules, they generate pure white-on-black characters.

Small full-color OLED screens are used in smartphones and on camera backs, but at the time of writing, large OLED screens are not a mature technology, partly because of production costs. A great variety of chemicals and layer configurations have been tried, and the application of pixels to a substrate has been attempted with vacuum deposition through a shadow mask and with a system similar to inkjet printing. Pixels that emit red, green, and blue light have also been used. Filtered pixels have been used. One dominant process has not yet emerged.

Longevity and brightness have been problems. Where red, green, and blue diodes have been used, the different colors deteriorate at different rates. While the human eye tolerates an overall reduction in brightness, it does not tolerate a slight color shift caused by blue pixels, for instance, losing brightness more rapidly than red pixels.

Because OLED screens promise to be thinner, lighter, and brighter, and may eliminate the need for a fragile glass substrate, there is a strong incentive to develop this technology, which seems likely to gain dominance in the future.

OLED panels may also become a source of diffuse, shadowless room lighting or office lighting when practical problems have been solved and costs have fallen significantly.

transducer | 27

The term **transducer** is used here to describe a *noise-creating device* that is driven by external electronics. By comparison, an **audio indicator** (discussed in the next entry) contains its own internal electronics and only requires a DC power supply. Either of these components is often described as a *beeper* or *buzzer*.

A **speaker**, more properly termed a *loudspeaker*, is an electromagnetic transducer but is seldom described in those terms. It has a separate entry in this encyclopedia and is defined here as a *sound reproduction device* that is larger and more powerful than a typical transducer and has a more linear frequency response.

While piezoelectric transducers formerly used crystals, only the more modern piezoelectric type that uses a ceramic wafer will be considered here.

Some transducers convert sound into electricity, but these are categorized as *sensors*, and will be discussed in Volume 3. The only transducers discussed in this entry are those that convert electricity into sound.

OTHER RELATED COMPONENTS

- **audio indicator** (see Chapter 28)
- **headphone** (see Chapter 29)
- **speaker** (see Chapter 30)

What It Does

An audio **transducer** is a device that can create an *alert*. It requires an AC signal that is supplied by external electronics, and in its simplest form may be referred to as a *buzzer* or a *beeper*.

Audio alerts are used in microwave ovens, washer/dryers, automobiles, gasoline pumps, security devices, toys, phones, and many other consumer products. They are often used in conjunction with touch pads, to provide audio confirmation that a tactile switch has been pressed.

The schematic symbols in Figure 27-1 can be used to represent any kind of audio alert, including **indicators**, which contain their own electronics to generate a simple tone or series of tones. Type A is probably the most popular symbol. Types B and C often appear with the word "buzzer" printed beside them for clarification. D and E are really symbols for a **speaker**, but are often used for an alert. F is the symbol for a *crystal*, now sometimes used to indicate a piezoelectric noise maker. G specifically represents an electromagnetic transducer, but is seldom used.

How It Works

A circular diaphragm is glued at its edges inside a cylindrical plastic enclosure, usually measuring from around 0.5" to 1.5" in diameter. The enclosure is sealed at the bottom but has an opening at the top, so that sound can emerge from the upper side of the diaphragm without being partially cancelled by sound of opposite phase that is emitted from the underside of the diaphragm. The enclosure also amplifies the sound by reso-nating with it, in the same way that the body of a guitar or violin amplifies a note being played on the strings.

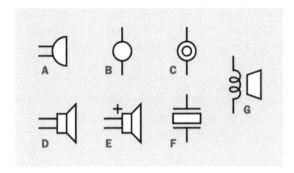

Figure 27-1. *An assortment of symbols which can repre-sent a transducer or an indicator. See text for details.*

The diaphragm is activated either electromag-netically or piezoelectrically, as described next.

Externally, a transducer may be indistinguisha-ble from an audio indicator such as the one pic-tured at Figure 28-1.

Variants

Electromagnetic

An electromagnetic transducer contains a dia-phragm that is usually made of plastic. Mounted on it is a smaller *ferromagnetic disc* that responds to the fluctuating field from AC passing through a coil. When the diaphragm vibrates, it creates pressure waves that are perceived by the human ear as sound.

A car horn is a particularly loud form of electro-magnetic transducer.

Piezoelectric

A piezoelectric transducer contains a diaphragm consisting of a thin brass disc on which is moun-ted a *ceramic wafer*. When an AC signal is applied between the piezoelectric wafer and the disc, the disc flexes at that frequency.

The term *piezo* is derived from the Greek *piezein*, which means "to squeeze or press."

Ultrasonic Transducer

The diaphragm in an *ultrasonic transducer* vi-brates at a frequency above the range of human hearing. This component may be electromag-netic, piezoelectric, or crystal-based. Often it is used in conjunction with an ultrasonic receiver as a distance measuring device. The two compo-nents can be sold pre-mounted on a breakout board. An output from the board can consist of a pulse train where the pulse duration is propor-tional to the distance between the transducer and the nearest sound-reflecting object.

An ultrasonic transducer is pictured in Figure 27-2. Its internal components are shown in Figure 27-3.

Figure 27-2. *The exterior of an ultrasonic transducer.*

Submersible ultrasonic transducers may be used in cleaning systems, where they agitate a liquid that dislodges dirt or debris. Ultrasonic trans-ducers are also used in echo-sounding and sonar equipment with marine applications.

Figure 27-3. *Inside an ultrasonic transducer, a small aluminum cone is the sound radiating element. The white blobs are adhesive to secure the thin wires.*

Formats

Some transducers are available in surface-mount format, measuring about 0.5" square or less. Because the resonant frequency is related to the size of the component, surface-mount transducers usually generate a high-pitched beep.

Values

Frequency Range

Audio frequency is measured in Hertz, abbreviated Hz, named after Heinrich Rudolf Hertz, the first scientist to prove the existence of electromagnetic waves. The H in Hz is capitalized because it refers to a real name. One thousand Hertz can be written as 1 kiloHertz, almost always abbreviated as 1kHz (note that the k is lowercase).

The human ear is often described as being able to detect sounds between 20Hz and 20kHz, although the ability to hear sounds above 15kHz is relatively unusual and diminishes naturally with age. Sensitivity to all frequencies can be impaired by long-term exposure to loud noise.

The most common frequencies applied to audio transducers range between 3kHz and 3.5kHz.

Piezoelectric elements are inefficient for generating sounds below 1kHz, but electromagnetic transducers are better able to generate lower frequencies. Their response curve can be approximately flat to frequencies as low as 100Hz.

Sound Pressure

Sound pressure can be measured in Newtons per square meter, often abbreviated as Pa. Newtons are units of force, while Pa is an abbreviation of Pascals.

The *sound pressure level* (SPL) of a sound is not the same as its sound pressure. SPL is a logarithmic value, to base 10, in units of decibels (dB), derived from the pressure of a sound wave relative to an arbitrary reference value, which is 20 micro-Pascals (20µPa). This is the agreed minimum threshold of human hearing, comparable to a mosquito at a distance of three meters. It is assigned the level of 0dB.

Because the decibel scale is logarithmic, a linear increase in the decibel level of a sound does not correspond with a linear increase in actual sound pressure:

- For each additional 6dB in the SPL, the actual sound pressure approximately doubles.
- For each additional 20dB in the SPL, the actual sound pressure is multipled by 10.

Bearing in mind that 0dB corresponds with the reference sound pressure of 20µPa, an SPL of 20dB represents a sound pressure of 200µPa (that is 0.0002Pa), and so on.

Many tables show an estimated decibel level for various noise sources. Unfortunately, these tables may contradict each other, or may fail to mention the distance at which a sound is measured. Figure 27-4 shows estimates derived by averaging eight similar tables. It should be viewed as an approximate guide.

Decibels	Noise Example
140	Jet engine at 50 meters
130	Threshold of pain
120	Loud rock concert
110	Automobile horn at 1 meter
100	Jackhammer at 1 meter
90	Propeller plane 300 meters above
80	Freight train at 15 meters
70	Vacuum cleaner
60	Business office
50	Conversation
40	Library
30	Quiet bedroom
20	Leaves rustling
10	Calm breathing at 1 meter
0	Auditory threshold

Figure 27-4. *Approximate decibel values for some sound sources (averaged from a selection of eight similar charts).*

Sometimes the claim is made that an increase of +10 on the decibel scale will correspond with a subjective experience that the noise is "twice as loud." Unfortunately, this statement cannot be quantified.

Weighted Sound Values

Subjective assessment of sound is complicated by the nonlinear frequency response of the human ear, which causes some frequencies to seem "louder" than others, even though their sound pressure is the same. The frequency weighting of the ear can be determined by playing a reference tone of 1kHz at 20dB and then doing an A-B comparison with a secondary tone at another frequency, asking the subject to adjust the gain of the secondary tone up or down until the two tones seem equally loud.

This procedure is performed for a range of frequencies. The test is then repeated with a louder 1kHz reference tone, at 30dB. Repetitions continue to a final reference tone of 90dB.

The resulting curves are known as *equal-loudness contours.* An averaged set, from multiple sources, has become an international standard with ISO number 226:2003. The curves shown in Figure 27-5 are derived from that standard. The curves show that the sound pressure of lower frequencies must be boosted by a significant amount to sound as loud as a 1kHz frequency, while a frequency around 3kHz must be reduced slightly, because it tends to sound louder than all others.

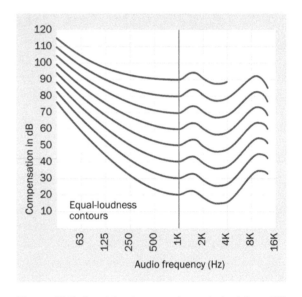

Figure 27-5. *Equal loudness contours derived from ISO 226:2003. See text for details.*

Although the accuracy of equal-loudness contours is controversial, they have been the basis of a widely used weighting system to adjust dB values to represent subjective perceptions of loudness. This *A-weighting* system remains the best-known and most widely applied audio standard in the United States, even though it has been criticized for assigning too little value to sounds that are brief in duration. If a sound level is expressed in dBA, it is A-weighted, meaning that the sounds to which the ear is least sensitive are assigned a value that is lower than their measured value. Thus, a tone of 100Hz has a dBA value about 20dB lower than its dB value, because the human ear is relatively insensitive to low-pitched sounds. dBA values are used in regulations that limit noise in the work place and other environments.

Unweighted Values

If sound intensity is expressed in dBSPL, it is a measurement of the actual Sound Pressure Level and has not been adjusted with the A-weighting system. A graph of unadjusted dbSPL values will display low frequencies as if they are more intense than the ear will perceive. In practical terms, subjective perception of low-end rolloff will be even more severe than the graph makes it appear.

If sound intensity is expressed merely in dB, probably it is unweighted and should be considered as dBSPL.

From a practical point of view, when choosing a tone for a transducer, a 500Hz tone may sound relatively mellow and not subjectively loud. A 3.5kHz tone can be a good attention-getting signal, as the ear is most sensitive in that range.

Transducers generally have a sound pressure rating in dBSPL ranging between 65dBSPL to 95dBSPL, with just a few products that can make more or less noise.

Measurement Location

The sound pressure from an audio alert will naturally diminish if the measurement point moves farther away. Therefore, any rating in decibels should be expressed with reference to the distance at which the measurement is made.

Measurement locations may be expressed in centimeters or inches, and may vary from 10cm to 1 meter, even in datasheets for different devices from the same manufacturer. If the measurement distance doubles, the SPL diminishes by approximately 6dB.

Limitations

A piezoelectric transducer is not intended as a sound reproducer, and does not have a smooth or flat frequency response. The curve for the Mal-

lory PT-2040PQ is not unusual, reproduced in Figure 27-6. This component measures about 3/4" in diameter, is rated for 5VDC, and uses only 1.5mA to generate 90dB (measured at a distance of 10cm). Like many piezoelectric audio devices, its response peaks around 3500kHz and diminishes above and below that value, especially toward the low end. While it is perfectly adequate as a "beeper," it will not reproduce music successfully.

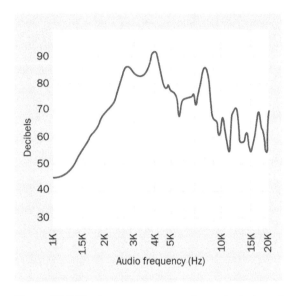

Figure 27-6. *The frequency response from a typical small piezoelectric transducer.*

An electromagnetic transducer is better able to generate low frequencies than a piezoelectric transducer. It has a low impedance that may be suitable in some circuits. However, it is slightly heavier than a comparable piezoelectric transducer, uses much more power, and as an AC device containing a coil, it can create electromagetic interference or may cause fluctuations in the circuit as an inductive load. It is also vulnerable to magnetic interference from elsewhere, while a piezoelectric transducer is not.

While an electromagnetic transducer can be used to reproduce speech or music, and will do a better job than a piezoelectric transducer, its performance will still be dissatisfying. A miniature speaker is more appropriate for the task.

Voltage

Transducers are typically designed to work with voltages ranging from 5VAC to 24VAC. The ceramic wafer in a piezoelectric transducer usually cannot withstand voltages much above 40VAC, and its sound output will not increase significantly above 30VAC.

Current

Typical piezoelectric transducers use less than 10mA and generate negligible heat. An electromagnetic transducer may draw as much as 60mA.

How to Use It

Appropriate Sound Intensity

An alert should be chosen with reference to the environment in which it will be used. To be easily heard, it should be at least 10 dB louder than ambient background noise.

Volume Control

Sound pressure can be lowered by reducing the voltage. Because a transducer does not consume much current, a *trimmer* can serve as a volume control. Alternatively, a rotary switch with a set of fixed-value resistors can select preset sound values.

AC Supply

Although a transducer is an AC device, it is unlikely to be designed for voltage that fluctuates positively and negatively either side of a neutral value. Typically it is intended for voltage that fluctuates between 0V (ground) and the rated positive value of the power supply, and its pins, wires, or terminals are usually marked accordingly. If it has wire leads, the red lead should be connected to the more-positive side of the supply. If it has pins, the longer pin should be more positive.

The alternating signal for a transducer can be supplied by any simple oscillator or astable multivibrator circuit. For a given peak voltage, a square wave will generate a louder signal than a sinusoidal wave. A simple 555 timer circuit can be used, with a second monostable timer to limit the duration of the beep if necessary. An astable 555 can be used to test the transducer and select the audio frequency that sounds best.

Self-Drive Transducer Circuit

If a transducer has three wires or pins, it is probably a *self-drive* type. The datasheet may identify its inputs as M, G, and F, meaning Main, Ground, and Feedback. The Feedback terminal is connected with a section of the diaphragm which vibrates 180 degrees out of phase with the Main terminal. This facilitates a very simple external drive circuit, such as that in Figure 27-7, where the frequency is determined by the transducer's resonant frequency.

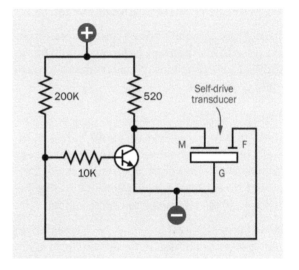

Figure 27-7. *A circuit to control a self-drive type of piezoelectric transducer.*

What Can Go Wrong

Overvoltage

Mallory Sonalerts, one of the largest producer of piezoelectric alerts, states that in the "vast majority" of returned products, the failure was caused by excessive voltage, often in the form of a transient voltage spike.

Leakage

If an alert makes a small, low-volume sound when it is supposed to be off, some current is leaking through it. Less than 1mA may be enough to cause this problem. According to one manufacturer, it can be fixed by placing a 30V Zener transient voltage suppressor diode in series with the alert, or by wiring a small **incandescent lamp** in parallel with the alert.

Note that when the alert is activated, the full supply voltage will be seen at the lamp.

Component Mounting Problems

Some alerts are packaged with mounting holes, but many are not. Those with pins can be soldered into a board, but those without must be glued in place or inserted into a cavity from which they cannot shake loose. Silicone adhesive is recommended, but care must be taken to avoid any of it dripping into the alert before it sets.

Moisture

If an alert will be used in a location where it is vulnerable to moisture, it should be of a type that is sealed against the environment. Even a sealed unit should ideally be oriented so that it faces slightly downward.

Transducer-Indicator Confusion

Externally, a transducer and an indicator often look identical, and some of them are not marked with a manufacturer's part number. Damage can be caused by applying DC to a transducer or AC to an indicator. If both types of parts are kept in inventory, they should be carefully labeled.

Connection with a Microcontroller

A piezoelectric transducer can be driven by a microcontroller, but an electromagnetic transducer is not appropriate in that role, because of its relatively higher current consumption and its behavior as an inductive load.

audio indicator | 28

An **audio indicator** is defined here as a noise-creation device that generates a simple tone or series of tones. Unlike a **transducer**, which requires an external source of AC to determine its audio frequency, an indicator contains its own electronics and requires only a DC power supply. Either of these components is often described as a *beeper* or *buzzer*.

While piezoelectric alerts formerly used crystals, only the more modern piezoelectric type that uses a ceramic wafer will be considered here.

OTHER RELATED COMPONENTS

- **transducer** (see Chapter 27)
- **headphone** (see Chapter 29)
- **speaker** (see Chapter 30)

What It Does

When DC power is applied to an **audio indicator**, in its simplest form it creates a continuous or intermittent tone of a fixed frequency. This is called an *alert*.

Audio alerts are used in microwave ovens, washer/dryers, automobiles, gasoline pumps, security devices, toys, phones, and many other consumer products. They are often applied with touch pads, to provide audio confirmation that a tactile switch has been pressed.

A few indicators are programmed to create a two-tone sound, or multiple-tone sequences.

See Figure 27-1 in the previous entry for an assortment of schematic symbols that may be used to represent either an alert or a transducer.

How It Works

A circular diaphragm is glued at its edges inside a cylindrical plastic enclosure, usually measuring from around 0.5" to 1.5" in diameter. The enclosure is sealed at the bottom but has a small hole at the top, so that sound can emerge from the upper side of the diaphragm without being partially cancelled by sound of opposite phase that is emitted from the underside of the diaphragm. The enclosure also contains electronics to generate one or more audio tones, and amplifies the sound by resonating with it, in the same way that the body of a guitar or violin amplifies a note being played on the strings.

A PUI XL453 piezoelectric audio indicator is pictured in Figure 28-1, fully assembled on the right, and with its circuit board and diaphragm removed on the left. This indicator creates a pulsed tone at 3.5kHz with a sound pressure of 96dB. It draws 6mA at 12VDC and measures approximately 1" in diameter.

For more information about the measurement of sound frequency and pressure, see "Frequency Range" on page 251 and "Sound Pressure" on page 251 in the previous entry.

Figure 28-1. *A typical piezoelectric audio indicator.*

Externally, an audio indicator may be indistinguishable from a transducer. However, internally, an indicator is almost always a piezoelectric device, in which a *ceramic wafer* is mounted on a thin brass diaphragm. The term *piezo* is derived from the Greek *piezein*, which means to squeeze or press.

A transducer (described in the previous entry) is a piezoelectric or electromagnetic alert that does not usually contain its own circuitry and must be driven by an external source of AC, which establishes the audio frequency.

The distinction between an indicator and a transducer is often unclear in parts catalogs, where all alerts may be identified as *buzzers*, even though they mostly beep rather than buzz.

Audio Frequency

For a discussion of audio frequency, see "Frequency Range" on page 251 in the previous entry.

History

Probably the earliest form of electrically activated alert was the door bell, in which a 6VDC battery-powered solenoid pulled a spring-loaded lever terminating in a small hammer. The hammer struck the bell, but the motion of the

lever also opened a pair of contacts, cutting off power to the solenoid. The lever sprang back to its rest position, which closed the contacts and repeated the cycle so long as power was supplied through an external pushbutton.

Subsequent systems used a small loudspeaker powered by AC house current through a step-down transformer. This created a buzzing sound and may have been the origin of the term "buzzer."

Small components that made a beeping sound only became common when digital equipment required a simple, cheap way to confirm user input or attract attention to the status of a device.

Variants

Sound Patterns

Because an audio indicator contains its own electronics, the manufacturer has the freedom to create various patterns of sound output.

The default is a steady tone. Other common variants include an intermittent tone and a dual tone that fluctuates rapidly between two frequencies. This is sometimes referred to as a *siren*. A few variants can generate an output pattern consisting of several tones in sequence, or effects such as warbling or whooping sounds, which are used mainly in alarm systems.

Formats

Some audio indicators are available in surface-mount format, measuring 1/2" square or less. Because the resonant frequency is related to the size of the component, surface-mount alerts usually make a high-pitched beep.

Panel-mount and board-mount formats range from about 1/2" to 1.5" in diameter. A small audio alert designed to be mounted on a circuit board is shown in Figure 28-2, with its top removed on the right to expose the brass diaphragm glued around the edges. The same component is shown with its plastic enclosure removed completely in Figure 28-3.

Figure 28-2. *An audio indicator approximately 0.5" in diameter, partially disassembled on the right, revealing its brass diaphragm.*

Figure 28-3. *The same indicator from the previous photograph, with its enclosure completely removed.*

Values

For an explanation and discussion of sound pressure and its measurement in decibels, see "Sound Pressure" on page 251 in the previous entry.

Audio indicators generally have a sound pressure rating in dBSPL ranging between 65dBSPL to 95dBSPL, with just a few products that make more or less noise. At 120dB and above, most products are packaged as alarm sirens ready for installation, often with a small horn attached. Their power consumption can be 200mA or more, and they are many times the price of a simple indicator designed for circuit-board mounting.

Voltage

An audio indicator containing its own electronics will almost always be rated somewhere in the range from 5VDC to 24VDC. Sirens intended for use with burglar alarms are often designed for 12VDC or 24VDC, as these are popular values for security systems with battery backup. However, in addition to a rated voltage, a datasheet may specify a wide range of acceptable operating voltages. For example, an indicator with a *rated* voltage of 12VDC may have an *operating* voltage of 3VDC to 24VDC. Naturally, the sound intensity will vary with the voltage, but not as much as one might assume. The graph at Figure 28-4 shows that the sound output from an alarm, measured in decibels, increases by only 8dB when voltage increases by almost a factor of five. Of course, the decibel scale is not linear, but human perception of sound is not linear, either.

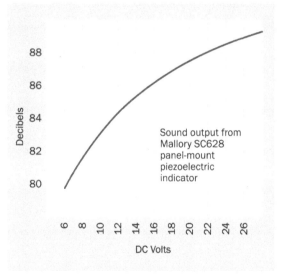

Figure 28-4. *Variation of sound output relative to voltage, in a commonly used piezoelectric indicator.*

Current

Typical piezoelectric indicators use less than 10mA (often as little as 5mA) and generate negligible heat.

Frequency

The most common frequencies for indicators range between 3kHz and 3.5kHz. Piezoelectric elements are inefficient for generating sounds below 1kHz.

Duty Cycle

Piezoelectric alerts generate very little heat and can be run on a 100% duty cycle.

If an alert will be pulsed briefly, the minimum pulse time is 50ms. A shorter duration will merely generate a clicking sound.

How to Use It

Appropriate Sound Intensity

An indicator should be chosen with reference to the environment in which it will be used. To be easily heard, it should be at least 10 dB louder than the ambient background noise.

Volume Control

Sound intensity can be reduced by reducing the voltage. Because an indicator does not consume much current, a trimmer can serve as a volume control. Alternatively, a rotary switch with a set of fixed-value resistors can select preset sound values.

However, in many indicators, variations in voltage may have relatively little effect on sound output, as shown in Figure 28-4.

Wiring

An indicator requires DC voltage. Because the indicator contains a transistor, polarity of the power supply is important. If the indicator has leads attached, the one intended for connection to the positive side of the power supply will be red. If it has pins, the longer pin will be for the positive connection.

What Can Go Wrong

The potential problems in an indicator are the same as those for a transducer. See "What Can Go Wrong" on page 254 in the previous entry.

headphone 29

The term **headphone** is used here to include almost any device that fits into or over the ear for the purpose of sound reproduction. (*Hearing aids* are not included.) Because headphones are used in pairs, the term is usually pluralized.

The term *phones* is fairly common as a colloquial diminution of *headphones* but is not used here.

An *earphone* used to be a single sound reproduction device designed for insertion into the ear, but has become rare. Pairs of *earbuds* are now common.

Because this encyclopedia assigns more emphasis to electronic components than to consumer products, this entry provides only a superficial overview of fully assembled headphones, and deals more with the drivers inside them, their principles of operation, and the general topic of sound reproduction.

OTHER RELATED COMPONENTS

- **transducer** (see Chapter 27)
- **speaker** (see Chapter 30)

What It Does

A headphone converts fluctuations of an electric signal into pressure waves that the human ear perceives as sound. It can be used for reproduction of music for entertainment purposes, or for speech in telecommunications, broadcasting, and audio recording.

Two symbols for headphones are shown in Figure 29-1. The symbol on the left shows a single headphone or earphone; when this symbol is flipped horizontally, it can represent a microphone. The pictographic symbol on the right has been used for many decades, but is still often found in schematics.

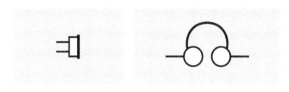

Figure 29-1. *Schematic symbols for a single earphone or headphone (left) and a pair of headphones (right).*

How It Works

Audio Basics

Sound is transmitted as pressure waves through a medium that is usually air but can be a gas, fluid, or solid. The speed of transmission will vary with the density and other attributes of the medium. Small hairs in the inner ears, known as *cilia*, vibrate in sympathy with pressure waves and transmit nerve impulses to the brain, which interprets the impulses as sound.

Three quantities describe the propagation of any type of wave, including a sound wave: its frequency (customarily represented with letter f), its speed of propagation (represented with letter v, for velocity), and its wavelength from peak-to-peak (represented by the Greek letter lambda, which appears as this λ symbol).

The relationship is defined by a very simple equation:

$$v = \lambda * f$$

Velocity is usually measured in meters per second, wavelength in meters, and frequency in Hertz, abbreviated Hz. One cycle per second is 1Hz. The H is always capitalized, as it refers to the name of Heinrich Rudolf Hertz, the first scientist to prove the existence of electromagnetic waves. One thousand Hertz can be written as 1 kilo-Hertz, almost always abbreviated as 1kHz (note that the k is lowercase).

The human ear is often described as being able to detect sounds between 20Hz and 20kHz, although the ability to hear sounds above 15kHz is relatively unusual and diminishes naturally with age. Sensitivity to all frequencies can be impaired by long-term exposure to loud noise.

Naturally occurring sounds can be converted to fluctuations in voltage by a **microphone**, which will be found listed as a *sensor* in Volume 3 of this encyclopedia. Artificial sounds can be generated as voltage fluctuations by oscillators and other electronic circuits. In either case, the output fluctuations can range between an upper limit set by a positive supply voltage and a lower limit established by electrical ground (which is assumed to be 0 volts). Alternatively, the fluctuations can range between the positive supply voltage and an equal and opposite negative supply voltage, with 0V lying midway between the two. This option can be less convenient electrically but is a more direct representation of sound, because sound waves fluctuate above and below ambient air pressure, which can be considered analogous to a ground state.

The concept of positive and negative sound waves is illustrated in Figure 29-2 (originally published in the book *Make: More Electronics*).

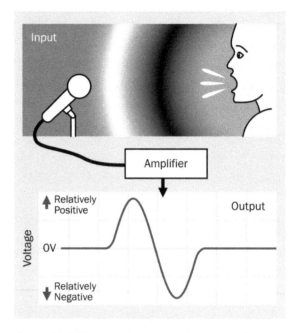

Figure 29-2. *The use of positive and negative voltages to represent a wave of high sound pressure followed by a trough of lower pressure.*

The topic of sound amplification is explored in detail in the entry on *op-amps* in Chapter 7.

A headphone inverts the function of a microphone by converting electricity back into air-pressure waves. This is done electromagnetically (moving a diaphragm in response to an electromagnet) or electrostatically (moving a membrane in response to electrostatic force between two charged electrodes).

Variants

Moving Coil

The most enduringly popular type of headphone uses a coil attached to a *diaphragm*. This is known as a *moving-coil* headphone, as the coil moves with the diaphragm. It can also be referred to as having a *dynamic driver* or *dynamic transducer*, "dynamic" referring to the movement of the coil.

The moving-coil concept is illustrated in Figure 29-3. The coil slides into a deep, narrow, circular slot in a magnet that is attached to the plastic frame of the headphone. The diaphragm is supported at its edges by a flexible rim. Variations in current passing through the coil create a fluctuating magnetic field that interacts with the field of the fixed magnet, causing the diaphragm to move in and out. A very similar configuration is used in many **loudspeakers**. Detail modifications may be made to increase efficiency, reduce production costs, or enhance sound quality, but the principle remains the same.

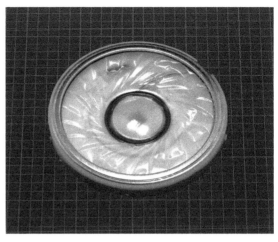

Figure 29-4. *The sound reproducing element removed from a headphone.*

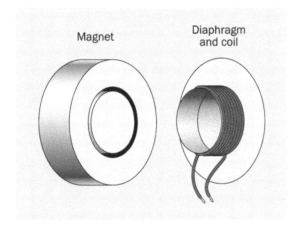

Figure 29-3. *The basic elements of a moving-coil headphone.*

The internal element of a headphone is shown in Figure 29-4. A plastic diaphragm is visible, measuring slightly less than 2" in diameter. The magnet and coil are concealed underneath.

The element in the previous figure is normally enclosed in an assembly such as the one in Figure 29-5, which incorporates a soft padded rim to rest upon the ear.

In an effort to achieve a more balanced frequency response, some designs use two moving-coil drivers in each headphone, optimized for low frequencies and high frequencies, respectively.

Earbuds, described after the next section, often use a miniaturized version of the moving-coil design.

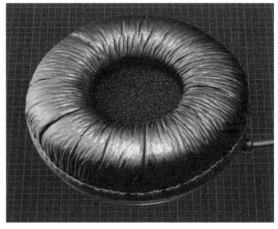

Figure 29-5. *The sound reproducing element from the previous figure is normally packaged inside this enclosure.*

Other Types

Electrostatic headphones use a thin, flat diaphragm suspended between two grids that function as electrodes. A fluctuating potential between the grids, coupled with a reverse-phase voltage on the diaphragm, will vibrate the diaphragm, generating pressure waves. A relatively high voltage is necessary to achieve this, anywhere from 100V to 1,000V, supplied through a conversion unit between the headphones and an amplifier. Electrostatic headphones are known for low distortion and an excellent high frequency response, at some extra cost.

Electret headphones work on a similar principle, except that the membrane is permanently charged, and a high voltage is not required. Electret headphones tend to be small, inexpensive, and not of high sound quality.

A *balanced armature* design, often referred to by the acronym *BA*, uses a pivoting magnet that is claimed to increase efficiency while reducing stress on a diaphragm. BA drivers can be extremely compact, contained within a sealed metal enclosure measuring less than 10mm x 10mm x 5mm. They are commonly used in conjunction with *in-ear* earphones, described in the next section.

Mechanical Design

Circumaural headphones use large soft pads to encircle the ear and block external noise. Their size tends to make them heavy, requiring a well-designed headband to provide comfortable support. *Supra-aural* headphones are smaller and lighter, resting on the ears instead of enclosing them. They cannot exclude ambient noise, and may have inferior bass response compared with the circumaural type.

Open-back headphones, also known as *acoustically transparent*, are favored by some audiophiles because their vented outer surfaces are thought to create a more natural sound, similar to that of a **speaker**. The open backs naturally allow ambient noise to intrude, but also allow the sound generated by the headphones to be heard by others in a room. *Closed-back* headphones contain their sound and provide more insulation against ambient noise.

Earbuds rest just within the outer folds of the ear, facing inward like a pair of tiny **speakers**. They are easily dislodged and provide very little insulation against ambient noise. Their use became common after the introduction of Apple's iPod. A pair of earbuds, one of them with its plastic cover removed, is shown in Figure 29-6.

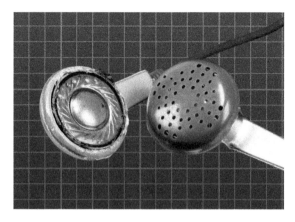

Figure 29-6. *A pair of earbuds, one with its cover removed to show the sound reproducing element, which closely resembles the diaphragm in a full-sized headphone.*

In-ear headphones are designed for insertion into the ear canal, often using a soft sheath that conforms with the ear like an earplug. This sheath is disposable for hygienic reasons, and because it may lose some of its plasticity with use. It excludes most ambient noise, and by minimizing the air gap between the driver of the headphone and the ear drum enables a high quality of sound reproduction.

In-ear headphones are also known as *in-ear monitors*, *IEMs*, *ear canal headphones*, *earphones*, and *canalphones*. A pair of in-ear headphones is shown in Figure 29-7, one of them with its foam sheath removed. The rectangular silver-colored object in the headphone on the left contains a transducer to create sound pressure.

A *headset* consists of one or two headphones plus a flexible microphone that extends to the promiximity of the mouth of the user.

Noise-cancelling headphones, popularized by Bose, monitor external noise with a built-in microphone and generate sound of opposite phase, to provide some cancellation. They are particularly effective on jet aircraft, where background noise tends to be consistent.

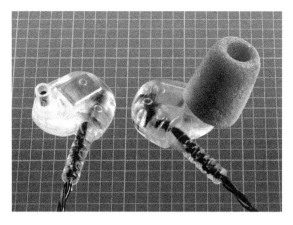

Figure 29-7. *A pair of in-ear headphones, supplied with disposable foam plugs that conform flexibly with the ear canal. The headphone on the left is shown with its plug removed.*

Although an *earphone* is almost obsolete, it is still obtainable from specialty suppliers. It has a high impedance, making it suitable for use with a *crystal-set radio*. An earphone is pictured in Figure 29-8.

Figure 29-8. *A vintage earphone of the type suitable for use with a crystal-set radio.*

Values

Intensity

Sound pressure is measured in decibels. For a complete explanation and discussion of weighted and unweighted decibel scales, see "Sound Pressure" on page 251 in the **transducer** entry.

Frequency Response

A plot of sound pressure as a function of frequency shows the *frequency response* of a headphone. Measuring the sound pressure meaningfully is a challenge, because the ear canal will add coloration to the sound and can amplify some frequencies while masking others. Ideally, measurement should be done at the ear drum, but this is not feasible. Consequently, high-end headphones are evaluated by making sound measurements inside simulated ear canals in a dummy human head.

A comparison between a high-quality $500 audio product and a transducer that is sold as a component for less than $1 illustrates the difference in frequency response; see Figure 29-9. The Sennheiser headphones have a smooth response that rises toward the low end, compensating for the lack of bass response that tends to be a problem in headphones, and the relative insensitivity of the human ear to low frequencies. The fluctuations at the high end are within about 5dB.

By comparison, the Kobitone emphasizes the range between 3kHz and 4kHz because its primary task is to be heard, and these are the frequencies where human hearing is most sensitive. Its low-frequency response trails off (although is still much better than that of a piezoelectric transducer, where the low response typically diminishes by 40dB to 50dB). The low-frequency output of the Kobitone is actually impressive bearing in mind that the component is only 9mm in diameter. It draws 60mA at 5VAC.

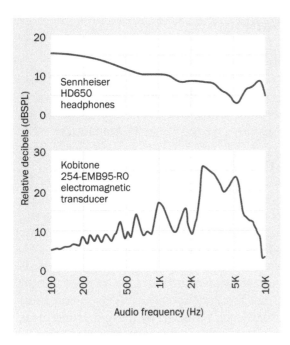

Figure 29-9. *Comparison between frequency responses of a $1 electromagnetic transducer intended as an audio alert and a $500 pair of headphones intended for sound reproduction. The upper graph is derived from a review online at headroom.com; the lower graph is from the manufacturer's datasheet.*

Some manufacturers of audio reproduction equipment prefer not to supply frequency response curves. Instead they may claim, for example, that the frequency response of a product ranges from 100Hz to 20kHz. This claim means very little unless it is accompanied by a range of sound pressure levels. If the frequency response is consistent within a range of, say, plus-or-minus 5dB, this may be acceptable. If the range is plus-or-minus 20dB, it is not acceptable. The ability to reproduce a high note or a low note is not useful if the sound is too faint to hear.

Distortion

The *total harmonic distortion* (THD) of any audio equipment measures its tendency to add spurious *harmonics* of a single frequency. If headphones are required to reproduce a pure 1kHz sinewave, they will also tend to create an additional 3kHz tone that is an artifact. This can be caused by the mechanical behavior of a vibrating diaphragm. The human ear recognizes distortion as a fuzzy or rasping sound. A square wave theoretically contains all the harmonics that are odd multiples of its fundamental frequency, and sounds extremely distorted.

THD should be less than 1% in good-quality audio devices.

Impedance

The electrical impedance of headphones is relevant in that it should match the output specification of the amplifier that drives them.

What Can Go Wrong

Overdriving

Headphones can be damaged by overdriving them. Because a low frequency requires larger excursions of a diaphragm to transmit the same energy as a high frequency, headphones are especially vulnerable to being damaged by bass at high volume.

Hearing Damage

Human hearing can be damaged by prolonged listening to headphones at a high volume. Some controversy remains regarding an acceptable limit for sound pressure.

Mismatched Impedance

If the impedance of headphones does not match the output of the amplifier driving them, distortion or a skewed frequency response can result. This is known as *mismatching*.

Incorrect Wiring

In most consumer products, a pair of headphones will share a common ground. While the connections in a typical three-layer jack plug have been standardized, hand-wired repairs or extensions should be tested carefully. Incorrect wiring will cause unpredictable results.

speaker | 30

The term **speaker** is a diminution of *loudspeaker*. The full word is now so rarely used, some catalogs do not recognize it as a search term. This encyclopedia acknowledges contemporary usage by using **speaker** rather than *loudspeaker*.

A fully assembled consumer product can be referred to as a "speaker," but it also contains one or more individual components that are described as "speakers." To resolve this ambiguity, referring to the components as *drivers* is helpful, but this practice can create more ambiguity because other types of components are also called "drivers." The only real guide to the meaning of *speaker* is the context in which it is used.

For the purposes of this entry, a speaker is a *sound reproduction device*, distinguished from a typical electromagnetic **transducer** by being larger and more powerful, with a more linear frequency response. A transducer may be used as a *noise-creating device* to provide an alert, informing the user of the status of a piece of equipment. Because some speakers have been miniaturized for use in handheld products, they may be used as transducers, allowing some overlap between the two categories.

Because this encyclopedia assigns more emphasis to electronic components than to consumer products, this entry provides only a superficial overview of fully assembled speakers, and deals more with the drivers inside them, their principles of operation, and the general topic of sound reproduction.

OTHER RELATED COMPONENTS

- **headphone** (see Chapter 29)
- **transducer** (see Chapter 27)

What It Does

A speaker converts fluctuations of an electric signal into pressure waves that the human ear perceives as sound. It can be used for entertainment purposes or to provide information in the form of spoken words or distinctive sounds (as in the case of a miniature speaker in a cellular phone, playing a ring tone).

The internationally accepted schematic symbol for a speaker is shown in Figure 30-1.

Figure 30-1. *Only one symbol exists to represent a speaker. This is it.*

How It Works

For a summary of basic concepts and terminology relating to sound and its reproduction, see "Audio Basics" on page 261 in the previous entry.

Construction

A speaker contains a *diaphragm* or *cone* with a coil attached to it. Fluctuations of current through the coil interact with a permanent magnet, causing the speaker to emit pressure waves that are proportionate with the current. The design is similar in concept to that of a headphone, shown diagrammatically in Figure 29-3. The primary difference is that a speaker of around 2" or more will use a cone rather than a flat diaphragm. The cone shape is more rigid and creates a more directional sound.

A 2" speaker rated for 1/4W with a 63Ω coil is shown in Figure 30-2, undamaged on the left but with its cone cut out on the right. The neck of the cone, which is normally inserted in the circular groove in the speaker magnet, is shown with the inductive coil wrapped around it.

Figure 30-2. *On the left is a 2" speaker. On the right, its cone has been cut away to reveal the magnet, with a circular groove in it. The neck of the cone, which normally slides into the groove, is shown removed.*

A speaker with a cone 4" in diameter is shown from the rear in Figure 30-3.

A miniature surface-mount speaker is shown from the front and from the rear in Figures 30-4 and 30-5. It measures just under 0.4" diameter and was made for Motorola. Its power rating is 50mW.

Figure 30-3. *The back side of a speaker with a cone measuring approximately 4" in diameter. Its magnet is the large round section that is uppermost. This unit is rated for 4W and has an impedance of 8Ω.*

A speaker designed for a cellular phone is shown in Figure 30-6. Note the close resemblance in design to the driver used in an earbud, shown in Figure 29-6 in the previous entry.

In the past, speaker cones were made from tough, fibrous paper. Modern cones are more likely to be plastic, especially in small sizes.

Figure 30-4. *Front view of a surface-mount speaker measuring less than 0.4" diameter.*

Figure 30-5. *Rear view of the speaker shown in the previous figure.*

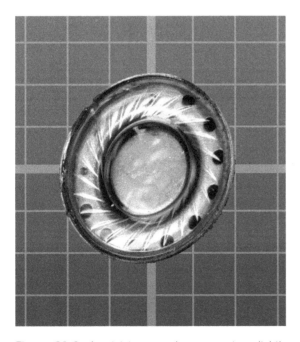

Figure 30-6. *A miniature speaker measuring slightly more than 1/2" diameter and only 0.13" thick, designed for use in a cellular phone. It has an impedance of 150Ω.*

Multiple Drivers

Generally speaking, a large-diameter speaker cone is more effective than a small cone at moving the greater volumes of air associated with reproduction of bass notes. However, the inertia of a large cone impairs its ability to vibrate at high frequencies.

To address this problem, a large speaker and a small speaker often share a single enclosure. A *crossover network* using coils and capacitors prevents low frequencies from reaching the small speaker and high frequencies from reaching the large speaker. The basic principle is shown in the simplified schematic in Figure 30-7.

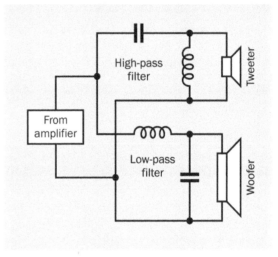

Figure 30-7. *The basic principle of a crossover network.*

Because the crossover network must be "tuned" to match the characteristics of the speakers, and because the combined sound pressure of the speakers must be relatively consistent over a wide range of frequencies, an actual network usually includes additional components.

Because the audio output from an amplifier consists of alternating current, polarized capacitors cannot be used. Polyester capacitors are common.

The small speaker in a pair is known as a *tweeter* while the large speaker is a *woofer*. Although these appellations were whimsically coined, they have endured.

More than two speakers may be combined in an enclosure, in a wide variety of configurations.

Venting

A speaker radiates pressure waves from the back side of its cone as well as from the front, and because the waves from front and back are opposite in phase, they will tend to cancel each other out.

In a miniature speaker, this problem can be addressed simply by sealing the section of the enclosure at the rear. For larger components, a more efficient enclosure can be designed with a *vent* or *reflex port* at the front. Pressure waves from the back of the speaker are diverted over a sufficient distance inside the cabinet so that by the time they emerge through the port, they are approximately in phase with low frequencies from the front of the speaker, although the back wave will lag the front wave by one wavelength.

This design is referred to as a *bass-reflex enclosure*, and was almost universal in high-fidelity components until amplifiers became increasingly powerful during the 1960s. At that point, Acoustic Research, located in Massachusetts, marketed a product line in which speaker enclosures were sealed, the argument being that when an amplifier can deliver 100W per channel, efficiency is no longer an issue, and a sealed enclosure can eliminate compromises associated with a bass-reflex design.

Acoustic Research referred to their concept as "air suspension," as the cushion of air in the sealed cavity helped to protect the speaker by limiting its excursion. This configuration is now often referred to as a *closed-box speaker*. Some audiophiles argue that it must always be inherently superior to a bass-reflex design, partly because of the one-wavelength lag time associated with a reflex port. However, as in many aspects of sound reproduction, the debate is inconclusive.

Resonance

The enclosure for a speaker will tend to have a dominant resonant frequency. This should be lower than the lowest frequency that the speaker will reproduce; otherwise, the resonance will emphasize some frequencies relative to others, creating unwanted peaks in the response.

One reason why high-quality speakers tend to be physically heavy is to reduce their resonant frequency. A modern Thiel speaker assembly, for instance, uses a front panel of particle board that is a full 2" thick. However, heavy enclosures are expensive to transport and inconvenient to locate or relocate in the home.

To address this problem, a tweeter and a woofer can be mounted in separate boxes. The enclosure for the tweeter can be very small, lightweight, and suitable for placement on a shelf, while the heavy box for the woofer can go on the floor. Human senses have difficulty locating the source of low-frequency sound, so the woofer can be located almost anywhere in a room. In fact, its single speaker can serve both stereo channels.

This configuration has become the default for computer speakers. It is also used in home-theater systems, where the woofer has now become a subwoofer capable of very low frequency reproduction.

Miniature Speakers

If an electronics project has an audio output, and the circuit board will be sharing an enclosure with a small speaker, the size of the box and the material from which it is fabricated will affect the sound quality significantly. A box made from thin hardwood may add resonance that sounds pleasing, if the speaker is being used just for simple electronic tones. By comparison, a metal box may sound "tinny." A box fabricated from a plastic such as ABS will be relatively neutral, provided the plastic is reasonably thick (1/4" being preferable).

Variants

Electrostatic Speaker

The principle of an electrostatic speaker is the same as that of an electrostatic headphone. A charged membrane is stretched between two grids in front of it and behind it that act as electrodes. Because the membrane is so light, it responds with very little latency, and its large surface area creates a diffused sound that many audiophiles find pleasing. However, a high voltage is required to drive electrostatic speakers, and they are not cheap.

Powered Speakers

A unit containing its own driver electronics is referred to as a *powered speaker*, and is used almost universally with desktop computers, because the computer itself does not contain a power amplifier. Powered speakers may also enable a more versatile crossover network.

A subwoofer may have its own amplifier allowing control of the cutoff frequency above which the speaker will not attempt to reproduce sound. The electronics can include protection for a speaker against being overdriven.

Wireless Speakers

A wireless link between a stereo receiver and its speakers will eliminate the speaker wires that are normally necessary. However, the speakers themselves must be powered, and will have to be wired to electric outlets.

Innovative Designs

The need for small speakers in consumer products such as laptop computers has encouraged innovative designs. The speaker in Figure 30-8 is just 1" square, and its shape is easier to accommodate in a small product than the traditional circular speaker. In Figure 30-9, the interior of the same speaker shows that inductive coils are applied to a square plastic diaphragm.

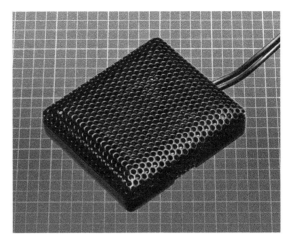

Figure 30-8. *A speaker 1" square, suitable for use with a small electronic device.*

Figure 30-9. *The speaker from the previous figure, opened to show its inductive coils applied to a square plastic diaphragm.*

Values

The typical *impedance* for speakers in audio systems is 8Ω. Small speakers may have a higher impedance, which can be useful when driving them from devices that have limited power, such as a TTL-type 555 timer.

In the United States, the diameter of a circular speaker is usually expressed in inches. Speakers larger than 12" are rare for domestic use. A 4" speaker used to be considered minimal because of its limited low-end frequency response, but much smaller speakers have become common in portable devices.

The low-end frequency response of a miniature loudspeaker designed to be surface-mounted on a circuit board will be very poor. The graph in Figure 30-10 was derived from data supplied by the manufacturer.

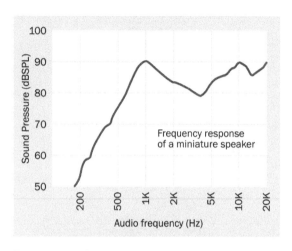

Figure 30-10. *Frequency response for a speaker measuring 15mm x 15mm x 5mm. Small dimensions and lack of an enclosure result in a negligible low-end response.*

Power rating for speakers is in watts, measured on a root-mean-square (RMS) basis.

Sensitivity is measured in decibels, at a distance of one meter, while a speaker is reproducing a single constant tone with a power input of 1W. A speaker assembly designed for undemanding home use may be rated at 85dB to 95dB.

Efficiency is a measure of sound power output divided by electrical power input. A value of 1% is typical.

What Can Go Wrong

Damage

As is the case with headphones, the most common problem affecting a speaker is damage caused by overdriving it. Because a low frequency requires larger excursions of a speaker cone to transmit an amount of energy comparable to that of a high frequency, loud bass notes can be hazardous to the cone. On the other hand, if an amplifier generates distortion (perhaps because it, too, is being overdriven), the harmonics created by the distortion can damage high-frequency speakers.

Magnetic Field

Even a small speaker (measuring 2" or less) contains a magnet that is sufficiently powerful to cause problems if it is in close proximity to other components, especially if sensors such as reed switches or Hall-effect switches are being used. Initial circuit testing should be done with the speaker as far away as possible, to eliminate it as a source of interference.

Vibration

Solder joints will be stressed if they are subjected to low-frequency vibration from a speaker. Loose parts may rattle, and parts that are bolted into place may become unbolted. The speaker itself may become loose. Thread-locker such as Loc-Tite should be applied to nuts before they are tightened.

Index

Symbols

A

NV (see nonvolatile memory)

O

octal counter, 134
OE (output-enable), 126
offset null adjustment, 56
OLED (organic light-emitting diode), 217, 227, 243, 246–247
on-resistance, 156
one-digit hexadecimal dot matrix, 236
op-amp, 39, 49–58, 90
    LM741, 49, 53
    calculating amplification, 53
    and comparators, 52
    confused input, 58
    controlling the gain, 53
    design, 49–52
    differences from comparator, 42
    dual inputs, 50–51
    as high-pass filter, 55
    as low-pass filter, 54
    negative feedback, 51–52
    offset null adjustment, 56
    oscillating output, 57
    potential problems, 57–58
    as relaxation oscillator, 55
    as single power source, 56
    use of, 53–57
    values, 52
    variants, 52
    what it does, 49
open collector, 40
open loop mode, 51
open loop operation, 42
open loop voltage gain, 53
open-back headphones, 264
open-drain outputs, 126
operating voltage, 259
operational amplifier (see op-amp)
optical maser, 197
optical switch, 34
optocoupler, 25, 33–37, 215
    D804, 37
    OPTEKD804, 34

analog output, 36
basic types, 36
bidirectional, 36
design, 34–34
function, 33–34
high speed, 36
high-linearity, 36
internal sensors, 35–36
logic-output, 36
potential problems, 37
schematic symbols, 34
use of, 36–37
values, 36
variants, 34–36
OR gate, 90, 91, 97, 98, 99, 100, 102, 103, 140, 142
organic LED (see OLED (organic light-emitting diode))
oscillating output, 47, 57
oscillator, relaxation, 12, 46, 55
output
    analog, 66
    clipped, 50
    descending, 136
    encoded, 134
    erratic, 48
    open-drain, 126
    oscillating, 47, 57
    parallel, 123
    push-pull, 42
    three-state, 142
    types of, 134
    ultraviolet, 222
    weighted, 134
output coupler, 198
output mode, comparator, 42
oven lamp, 176
overvoltage protection, 2, 6, 155

P

Pa (Pascals), 251
panel, 243
Panelescent electroluminescent lighting, 244
parallel input, 134
parallel-enable pin, 134

parallel-in, parallel-out (PIPO) shift converters, 125
parallel-in, serial-out (PISO) shift register, 124
parallel-serial converters, 124
parchmentized thread, 172
passive infrared motion detector, 211
passive matrix LCD, 161
phase angle, 7, 13
phase control, 2, 7–7, 13, 22
phones (see headphone)
phosphor, 192, 239, 244
phosphorescence, 244
photocell, 35
photodarlington, 34
photodiode, 26, 34
photon, 207
photoresistor, 34
phototransistor, 26, 34, 45
photovoltaic PIN diode, 35
piezein, 250
piezo, 258
piezoelectric alerts (see audio indicator)
piezoelectric transducer, 249, 250, 253
pin
    common, 161
    floating, 105
    inhibit, 153
    parallel-enable, 134
    reset, 70
    trigger, 69
pin base, 177
PIN diode, 35
pin identifier, 133
pip, 184
PIPO (see parallel-in, parallel-out (PIPO) shift converters)
pixel arrays, 231, 235–236
PMOLED (passive-matrix OLED), 246
PN junction, 207
PNPN device, 3
polling a keyboard, 127
positive feedback, 44
positive logic, 90

About the Authors

Charles Platt is a contributing editor and regular columnist for *Make:* magazine, where he writes about electronics. He is the author of the highly successful introductory hands-on book, *Make: Electronics*, and its sequel, *Make: More Electronics*. His science fiction novels are currently being reissued by Stairway Press.

Platt was a Senior Writer for *Wired* magazine and has written various computer books. As a prototype designer, he created semi-automated rapid cooling devices with medical applications, and air-deployable equipment for first responders. He was the sole author of four mathematical-graphics software packages, and has been fascinated by electronics since he put together a telephone answering machine from a tape recorder and military-surplus relays at age 15. He lives in a Northern Arizona wilderness area, where he has his own workshop for prototype fabrication and projects that he writes about for *Make:* magazine.

Fredrik Jansson is a physicist from Finland, with a PhD from Åbo Akademi University. He is currently living in the Netherlands, where he works on swarm robotics and simulates sea animals in the computational science group at the University of Amsterdam. Fredrik has always loved scavenging discarded household electronics for parts, and is a somewhat inactive radio amateur with the call sign OH1HSN. He also fact-checked Charles Platt's previous book, *Make: More Electronics*.

Colophon

The cover and body font is Myriad Pro, the heading font is Benton Sans, and the code font is Ubunto Mono.

CPSIA information can be obtained
at www.ICGtesting.com
Printed in the USA
BVHW020938050122
625528BV00017B/712